Real Water in Mirage

A. V. Bhagwat

authorHOUSE®

AuthorHouse™ UK Ltd.
500 Avebury Boulevard
Central Milton Keynes, MK9 2BE
www.authorhouse.co.uk
Phone: 08001974150

First published by AuthorHouse 5/27/2009

ISBN: 978-1-4389-2318-5 (sc)
ISBN: 978-1-4389-2319-2 (hc)

Library of Congress Control Number: 2008910188

About The Author

The author, Arvind Bhagwat is a graduate in three different subjects, namely, Science (BSc), Civil Engineering (BSc Engineering, Civil), and Indian Classical Music (B-Music), and retired in the year 2000 as Chief Engineer, Water Resources Department of state of Madhya Pradesh, in India. Philosophy was his much cherished subject right from when he was a student and he studied Indian and Western philosophy in his teen age and his twenties, and science and philosophy occupied his whenever available leisure all these days thereafter. He had written a very small book in Hindi language, on one particular aspect of Indian philosophy in 1980, which was locally published.

The Author's Observation

"In the year 1999 at Mumbai airport, I happened to read 'A brief History of Time' and I strongly felt that even scientific deliberations must not always be allowed to pass as truth, particularly when they include speculations rather than theories, for they only deal with phenomenal world as objectively observed, and without regard to human consciousness they try to find an unified theory (that should explain and predict even the human behavior), which must be a futile attempt."

A continued thinking on role of consciousness in the observed phenomena has resulted into the present book.

Author A.V Bhagwat dedicates this work to Bertrand Russell

Contents

Synopsis

Existent and non-existent although look adjectives meaning opposites, they are not entirely so. For a thing to be non-existent it has to be existent at some time in the past or at present, since we have to use present tense form of the verb 'to be' to express its non existence. A thing absolutely non-existent cannot be spoken of, since we cannot imagine such a thing at all.

Reality and falsehood is another pair of words which philosophers and scientists use very often to express what is in their minds, such as the Sun goes round the Earth is false and the other way round is the reality. Science although looks to have gone far ahead in search of realities of the phenomenal world but it is really in its childhood as physicist John Wheeler says that real big discoveries are yet to come. The most cherished Big-Bang model of origin of Universe is nothing but only a semi-logical hypothesis about how should the present configuration of stellar bodies have originated. The Universe consists of matter, energy, time, space and seeming causality and to assume existence of a point mass at incredibly high temperature and vast amount of energy to cruelly blast the thing into pieces, points in no way to the origin of the Universe.

Study of a phenomenon is not really to know how and why it happens but is to know how and why they are perceived to happen. If nothing in the Universe had any consciousness and intellect, the Universe would never have been known to exist. We cannot sever phenomenon from its perception.

The book deals with all connected aspects of phenomena and their perceptions and derives conclusions, which may be astounding to those who take the Universe to be a rigorous rigidity. The books written by men of science in the recent past leave the readers bewildered at the end for science has not yet reached the zenith nor has it fully fathomed deepest seeming uncertainties. Writing about black holes, singularities and Big-Bang introduces the readers to the recent advancement in physics but does not offer any satisfactory solution to the riddle of existence, for physicists themselves are not satisfied with the discoveries so far and that is why physics has not come to an end.

Physicists are now realizing role of consciousness in evaluating all phenomena but they are not sure how the two should be inter-related. Toeing with George Berkeley's philosophy, rejecting the existence of matter gives rise to equal number of problems as when being in line with the philosophy of some materialists, who take the matter to be everything and give only a trifling recognition to the conscious perception of it. Causes of perception when scanned to their utmost fineness show a great departure from generally accepted ideas and their center of action shifts much towards subjective-ness rather than thinking of a naive realist. A perceived mirage is far from its real outer causal ingredients. The Universe, which looks so very rigidly existent, may just fade away when the reasons of our conviction that it exists, are passed through finest sieves of reason.

This book satisfyingly solves riddles that arise when an observed phenomenon is not rightly connected to its conscious perception and shall present a new vision to look at the Universe as a whole.

Preface

While there are men who willingly or unwillingly take the world as they perceive and toil all their lives to collect wealth and comforts, there are a few others who cannot suppress the curiosity to know the how and why of the existence of the Universe, and spend all their lives in pursuit of knowledge and endeavor hard to reach some conclusion. And this has been true ever since the advent of human life on this revolving globe.

Great thinkers and philosophers have always been enlightening the world, be it in the East or the West. We cannot compare one with the other because the magnitude and altitude of their works has been so great and high that it seemed always difficult to find points of comparison for a layman almost stunned at their towering achievements. It is however to be noticed that although much different is the metaphysical logistics of thinkers of the West and the East, the formulation of conclusions for well being of mankind are not much in variance. Thinkers of the remote past perceived the world, analyzed its existence on metaphysical and psychological planes, blended their so drawn conclusions with fundamentals of human behavior and stipulated rules of life and ethics and became lighthouses for the bewildered man. Religions thus originated.

Let me declare in the first place that this book does not discuss any religion nor is its purpose to discuss it, and if any passing reference is made about anyone who has been the founder or supporter of any concept connected to any religion, it is only on metaphysical plane.

Although we know not of the first philosopher of the world but we cannot help concluding that he must have been the one who did not only see the world but tried to look beyond it. Indian scriptures, the Vedas, have been composed thousands of years ago and these huge works indicate how inquisitiveness of man has been adding new dimensions to itself till it transcended beyond the ordinary perception of the world and culminated into a sublimely crystallized hypothesis in the form of Upanishads.

The great philosophers of the West have added immensely to the world treasure of knowledge more especially by coming boldly out of old dogmas and then aligning themselves as far as necessary with the developing scientific concepts and at the same time maintaining their sharp speculatory approach. This has given us many philosopher scientists from Descartes to Stephen Hawking.

Questions like, when this Universe came into existence, where in the space it is located and if there has to be some reason for everything that happens, what reason was behind coming into being of this Universe?, have been so intelligently dealt with by thinkers of the West and the East that one marvels at their capacities to analyze or synthesize the basic ingredient ideas.

As for the philosophy every individual has the right to differ partly or fully with any proposition made by any other and that a well studied disagreement adds new heights to the knowledge is a fact. History bears this out.

Introduction

Assuming the things in this Universe must have been created one after the other, which of the three things must precede the other two? Space, time, or the matter? Obvious answer would be the space, for there has to be something to hold matter into it. If God created the Universe He must have created the space and with this material he must have proceeded to create matter or the time. But when God was all alone and there was no material with him to work upon, He must have created the space out of Him only, and then He must have created matter out of space itself and then He created element or a thought known as time out of him or out of space. Science today treats space and time not as two entities independent of each other but scientists are using space-time as one single variable, and are treating the difference of two events either space like or time like, depending on whether the value of a certain mathematical expression is positive or negative.

If we further assume the principle, that nothing can be done without a thought to do, to be applicable to the creation of the Universe, then we have to conclude that there must have been some super cosmic intelligence that must have willed things to happen, world to be created and stars and planets to rotate. If this is accepted then I have to make an astounding statement that intelligence does not need brains at all. Brain is a thing made of matter and matter did not exist at all when the great intelligence willed the creation of the Universe. We may call this formless, shapeless, matter less and timeless super intelligence as God if that is convenient for philosophical discussions; thus this cosmic Super Intelligence willed and everything came into being. The Holy Bible tells us that in the beginning God created the heaven and the Earth. God said "Let there be Light", and there was Light. God said "Let there be firmament in the midst of waters, and let it divide the waters from the waters" and there was firmament as commanded. How plain and simple. The thing, which we may add to this, is that God made everything out of Himself only since there was no other material except Him to work upon. If this is true then we have to accept that everything which we perceive is this Super Intelligence, only having taken different names and forms and this includes the perceiver also. It must be determined imagination of the Super Intelligence, which must have taken shape in the form of space, matter, time and energy. The idea is found in philosophy of the East and also in the West.

In dreams we create dreamland in a dream space. We create dream objects, and it must be our own imagination of objects, which soon takes shape. Thus we take the role of Super Cosmic Intelligence while creating our dreamland. There is nobody else to create our dreamland or to interfere with our dreams. If I dream, everything of dream including space, time, matter, energy and logic thereof is my creation. But here one thing has to be noticed and that is that 'I' who is participating in the dream as perceiver is not exactly the 'I' who is lying comfortably in his bed. The 'I' of dreams might be running or swimming or playing tennis in the dreamland. The sleeping 'I' is not identical with the dream perceiving 'I'. The dream perceiving 'I' is a creation just like other objects of the dream and the creator is the sleeping 'I'. Similarly the time in the dreams is not identical with that of the waking state. We may dream an event taking days or hours while we took a nap for few minutes only.

Time, space, matter, energy and the phenomenon we consider being causality or the chain of cause and effect, are the basic ingredients that constitute or result into this Universe. An examination of these fundamentals and their role is very necessary before arriving at any model aiming to depict origin of the Universe. The idea that there has to be a creator for everything that is created is derived from everyday experience of man who finds no daily usable things made without a maker. A box, a knife, a chair, a candle, a match, a gun or say bread, pudding, wine and the like have makers, and he applies this concept to animals, trees, rivers, oceans Earth, planets, the stars and galaxies equally and concludes existence of a maker. Thus it may be said that God is man's imagination. God is not directly perceived, since He has no visible form or shape, but is inferred from what He is supposed to have created. This is common view of the people not trained in scientific or philosophical studies.

However there are some more difficult questions regarding the creator and the creation. Space, time, matter and energy are the fundamental entities of which the Universe is made up of. The first question is about the order in which these things were brought into existence or whether they were created simultaneously. The second question would be why this creation is after all. What is the reason behind entire creation? Why didn't God choose to live by Himself without bothering to create and maintain such an unfathomable Universe? This question has its root in the fact that man has always found some cause for every happening in ordinary day to day life, and he extends this observation to the entire creation. The third question is whether the seen creation is real or is something like dreamland, having dream objects in the dream space with dream logic connecting and explaining dream phenomena. In the West the philosopher Descartes who is supposed to be the pioneer of modern philosophy, has raised such doubts and he maintains that he proves his own existence by doubting everything, and in turn he does not doubt his own doubts. Had man never dreamt, this question would never have cropped up. However we should be thankful to our capacity to dream which gave us the power to differentiate between real and unreal, facts and illusions, things and their images and truth and falsehood.

The book deals with finding answers to questions raised above not only from philosophical point of view but also by relying on discoveries made by physicists about perceptible Universe so far as they are logical explanation of physical phenomena and not simply an empirical relation between one happening and the other connected looking happening. If a trained deaf man interprets lip movements of the speaker empirically, he is still unaware of the voice, which really conveys the message fully through its modulations, accents, tone and the depth. This book concerns the voice more than the observed and lip movements, and presents a model aiming at resolving the riddle of the entire perceptible existence.

Chapter 1

The Matter

Matter, substance or body of some substance, or an object are common words we use in everyday life and mean thereby hundreds or thousands of things around us and also our limbs, heart, brain and the likes; but a philosopher or a scientist does not stop at the simple perception of the substance but inquires into what it is made of and wherefrom it has come and why does it exist at all. To the philosophers who think in terms of subject and predicate of a proposition, about anything defines substance as a subject of any proposition describing the substance. The mathematician philosopher Leibniz and others have gone so deep into defining substance that it cannot be anything else than what it is.

 In a proposition the attribute of a subject forms the predicate and substance has been defined as something which cannot be predicate in a proposition about the substance. Philosopher defines the substance as a thing, which does not depend on any other substance for its existence. Substances exist independently. Some philosophers believe that a substance or matter has also an action in it. If a substance exists this moment or this day and also next moment or tomorrow, it is said to possess the action of carrying its existence over to the next moment or the next day.

Scientists, whose thinking is analytical, define matter as something that occupies space, possesses weight and is perceptible to our sense organs. Although much has been discussed about space even to the extent whether space really exists or not, what scientists mean by space is what is understood by a layman that is some such entity which allows other objects to be situated in it. Weight is the force with which a body is attracted towards Earth or any other planet or star on which it is situated. About perception by our senses it shall be explained later on that there are only some of the properties that are capable of being subject matter of our sense organs and the mind and the complete substance in its real entirety is never perceived by us or any other living being. However the one line definition of matter proposed by scientists can be taken to be sufficient, when discussing the research made so far, to know what the matter is really made of and the branch of physics dealing with this is known as Particle Physics.

Scientists have analyzed the matter to such an extent that an atom of an element seems to consist of scores of different particles some stable and others unstable and the scientists are now grouping them into some definite categories so as to convey which is what. These particles cannot be touched or seen by naked eye but can be detected by elaborate experiments.

Initially it was discovered that an atom is the smallest particle of an element. By the end of nineteenth century another particle electron was detected and in 1932 it was found that an atom consists of a nucleus having protons and neutrons in it and electrons kept on orbiting round the nucleus same way like planets go round the Sun. The positive electric charge of the protons is balanced by the negatively charged electrons. The neutrons being neutral electrically, the atom as a result is neutral. An atom of hydrogen has one neutron and one proton at the center forming the nucleus around which only one electron orbits and since the amount of electric charge on the proton and the electron is exactly equal in magnitude but opposite in nature the atom as a whole is electrically neutral. An atom of helium has two protons and two electrons and when two hydrogen atoms fuse into one, an atom of helium is formed. This fusion is associated with release of tremendous amount of energy in the form of heat and light and this is what is happening continuously in the Sun and that is why Sun emits immense heat and light. The Sun has yet ample of its fuel that is hydrogen and it will last for millions of years to come. So no worry. An atom of carbon has 6 neutrons, 6 protons and to balance the positive charge of protons it has got 6 electrons orbiting about its nucleus. An atom of gold has 79 protons and 118 neutrons in its nucleus and has 79 electrons orbiting round it. It was also found that the number of electrons in an orbit is governed by what is called $2n^2$ law that is the maximum number of electron that an orbit can contain is equal to twice the square of the orbit number. Thus the first orbit can contain not more than two, second orbit eight the third orbit eighteen electrons and so on.

In 1932 another particle was discovered which had its mass equal to that of an electron but had an equal positive charge on it and was latter on named as positron. It was also noticed that when an electron and positron came into contact they annihilated each other radiating only energy like gamma rays. Thus positron was found to be antiparticle of electron. By 1975 it was found that an atom has got scores of elementary particles of varying masses and varying electric charges. Like positrons and electrons there are many such pairs of particles and their antiparticles, which annihilate each other, giving immense energy. The matter and energy are thus inter-convertible. Energy can be converted into pair of opposite type of particles. Further some particles convert themselves into some other particle and the process is known as decay. We are not sure as to whether the number of elementary particles to compose an atom is still found to increase over years to come since research still continues.

Science has gone far ahead in knowing how an atom is formed and of what elementary particles it is made of. Of course the questions like why electron or a proton should have an electric charge and why such a structure of atom should be as it is and above all why matter should come into existence at all, are yet to be answered and until now there is no progress made in this regard. Such questions were considered being almost out of area of science in the past, but recent inclination of the scientists towards what used to be philosophers field, poses a challenge to scientists themselves proving thereby that philosophy and science are almost inseparable.

When two particles of opposite nature, say an electron and positron collide they annihilate each other and energy in the form of gamma rays is emitted. Einstein has, by his famous equation $E=MC^2$, prescribed method to calculate energy so emitted. Here E stands for energy, M is the mass and C is the velocity of light which is 30000000000 centimeters per second. You may assess how big energy is released by destruction of one gram of mass. The energy thus is measured in Joules. But the question why the matter and the energy should be inter-convertible has not yet even been attempted to be answered by the scientists; it seems as though. If it be said that science attempts to know only 'How' of the things and the phenomena and question 'Why' is left to the philosophers to ponder upon, then as mentioned earlier, the modern trend of inquisitiveness of the scientists does not fully support this view. Physicists Stephen Hawking and others are attempting to know the Gods plans behind all this creation and this certainly goes to establish that science is not much different from philosophy although progress made by the scientists in answering

'Why' of the Universe is hardly tangible. Philosophers on this planet are ahead of scientists in attempting to answer this 'Why'. Science has to grow yet.

Returning to our discussion about elementary particles constituting an atom, it is said earlier that the number of such discovered particles is ever increasing. There are muons, which are 207 times as heavy as electrons, but having electric charge equal to that of electrons. As expected all such particles have antiparticles and when they collide with them there is release of energy and the pair of particles is wiped out from existence.

Scientists have now grouped elementary particles according to some similarities found in them. Particles named as Baryons resemble the protons and neutrons but have larger masses. Meson is a class of particles which include particles having varying masses from one seventh of a proton to those many times heavier than a neutron. The class Leptons includes electrons, muons and their companions like what are known as neutrinos. Quarks are particles, which constitute even proton and other negatively charged particles. It is said that proton is made of three quarks. As one might expect there are antiquarks too. These categories of particles have further been branded as up, down, strange, charm, bottom and top depending upon their properties and it is said that all matter is made up of six types of quarks and six leptons and their antiparticles also exist. It is however interesting to note that many such particles have a very short life. Thus a muon decays into an electron, a neutrino and its antiparticle in about half millionth part of a second. Some of these particles simply seem to pop into existence out of nothing and this is against all rules of logic known to this world. If even a small particle can come out of nothing why not extend the principle and say that entire Universe has come out of nothing? To circumvent this, physicists now say that entire space contains all kinds of such elementary particles dispersed all over it and that is from where atoms crop up and so on. However why the space should be plagued with these particles is yet unknown to science. Some physicists strongly believe that the space consists of fields of all such particles and it is from there that particles are found to pop up suddenly and into that they seem to go back.

And there is then what is popularly known as cosmic dance. Particles decaying into another particle or particles, one particle breaking into two and the two again joining to make the particle, and the virtual particles emanating from electrically charged particle causing itself to be absorbed by similar particle and showing or exerting a force of attraction or repulsion, is all that has been included in the words cosmic dance. Now those who know something about dance shall agree that dance is a very systematic and rhythmic movement of the body and its limbs, and any irregular movement or a movement without any purpose can not be termed as dance. Some think that the 'Tandav Nritya' a dance, full of actions and gestures, performed by Lord Shiva of Hindus represents this cosmic dance of subatomic particles, but this is only their afterthought when the cosmic dance was discovered or inferred by the physicists. For if this was not true, they would have been able to declare or at least predict accurately centuries ago the facts discovered in recent years about movements of the subatomic particles. Mythologies or religions can not be interpreted later on to match the hard earned scientific knowledge although philosophy can be. Philosophical knowledge is as hard earned as that of science although the methodology is quite different.

Scientists also claim to have detected presence of invisible matter termed as dark matter and it is guessed that the amount of such matter is much more than the visible matter or matter susceptible to our perception.... This speculation is from observing motion of some stellar bodies found deviated from their normally expected path of motion due to supposed gravity pull of some invisible masses. It thus appears that from tiniest particles which constitute an atom to the very large lumps of masses like stars, our knowledge is so incomplete that there appears much more to be known than what is known already, and I unhesitatingly agree with John Wheeler the physicist when he says 'the greatest discoveries in science are yet to be made'. The questions like why the matter is, how much exactly is matter and what is matter, are not yet fully answered. The question why matter is, lies much far away from the discoveries made so far in physics and chemistry although both claim to deal with matter.

Now let us examine how matter or a substance is perceived by us. The sense organs are eyes, ears, nose, tongue and sense of touch which extends all over our skin and whatever is perceived about a substance is through these five sense organs. A number of inventions have given us apparatuses to be used as extensions of our sense organs. Telephone is an extension of organ of hearing, Television, telescope or a microscope an extension of sense of sight. All perceptions are through natural organs or through their extensions. The picture of Universe depends upon efficiency of sense organs or their extensions. Man's eye cannot detect electromagnetic waves on either side of the spectrum of sunlight and the vision is limited only to violet, indigo, blue, green, yellow, orange and red colors, and to different combinations and shades of one or more of these colors. White color is combination of all these colors of a spectrum and the black is absence of all visible light. If we could see the ultraviolet radiations or infrared waves or if we were unable to detect any or more than one color, the world would look entirely different from what it looks now. Light travels in the form of transverse waves and the distance between two crests or two troughs is known as wavelength. Light waves which radiate violet color has wavelength of about 3900 AU. A.U stands for Angstrom units, which is equal to one ten millionth of a centimeter. Wavelength of red colored light is about 7500 A.U and those of all other colors lying between these two figures. Waves having wavelengths smaller than that of violet light are known as ultraviolet rays that include X rays and those beyond 7900 A.U are known as infrared waves which include heat rays and radio waves. If man could see infra red or ultraviolet radiations his visual world would be entirely different having colors we don't know of. It is said that dogs don't perceive spectral colors except white and absence of light i.e. black. Thus dog's world is then a black and white movie.

Man can hear sounds of frequencies lying between 20 and 20000 cycles per second. Frequency is the number of vibrations of source of sound in one second. Frequencies lower than 20 and higher than 20000 cycles per second are beyond man's range of hearing but dogs can hear still higher pitched sounds. Dog's audio world is having many more sounds than they are for us. Snakes do not have organ of hearing at all. What a relief they must be feeling to have ditched noise pollution. The world is silent for them. The snakes feel a vehicle passing by through vibrations transmitted through their long bellies in contact with Earth surface. One can imagine hundreds of such examples where properties of things look different to different observers. Horses can smell tigers from more than a mile while man never becomes aware of a tiger till the tiger leaps at him. This leads us to the conclusion that properties of a substance really depend on observer's capacity or his arrangement of sense organs. If our eye lenses were cylindrical we would see everything elongated. When one has jaundice white looks yellow. It cannot be logically claimed that present construction and arrangements of our eye or ear should be the right one. It could as well have been different. It is illogical to think that the world is as man sees or hears and is not as what dogs, horses, snakes or other animals see hear or smell. Many species have animals that outnumber human beings and thus no resort can be taken to principle of majority to prove that man's world is the only real world.

It is said that it took millions of years on this planet to develop intelligent life [like ours of course] Simply said, the intelligence means capacities to correlate signals sent by the sense organs to the brain and to draw simple or complex conclusions known as pieces of knowledge. It is easy to understand that if sense organs convey altogether different perception to the brain, the conclusion will also be different. If out of five sense organs one or more were denied to us our knowledge of the Universe would have been different from what it is now. Who knows after about a couple of billion years the present intelligent life may develop more number of sense organs or much more enhanced power of thinking thus changing the entire panorama of the perceived Universe.

Leaving aside any of such speculations for the moment we come to the question as to how our sense organs transmit the signal to our brains. The light coming from the object after refraction through our eye lens makes an inverted image of the object on our retina, and it is here that the optical phenomenon ends. No light wave ever reaches our brain. The signal passes from the retina to the brain through nerves by a sort of electrochemical change and there is absolutely no light involved between the retina and the brain. The brain has learned through habit that inverted image corresponds to an upright object and that it has such color and such a size. If eye receives a jerk a false signal is conveyed to cause brain to 'see' a light flash in absolutely dark room. This indicates that there need not be any light to cause the brain to 'see' the light. If we were not able to converge our eyes towards one point [convergently; squinting so to say] we would have seen two objects blurring each other by overlap. Frogs and toads do see different views by their pair of eyes, which are so situated that converging is not possible.

It follows from all above discussion that the observed properties of a substance do not really exist in the substance but are result of observer's arrangement of sense organs and his/her capacity to correlate impressions and draw conclusions, which form piece of knowledge about the substance. Philosophers of the West and the East have been pondering over the question since ages whether the existence of the external world is real or partly real or altogether unreal and since the observer himself is an object for another observer the question becomes still more complicated to resolve. Mathematician philosopher Descartes thought that there was no way to know whether impression transmitted by our sense organs is actually copy of the object or not. He adopted method of doubt in his philosophical investigations. The existence of the perceiving subject, he said was proved because he could doubt any proposition about any empirically existing thing or a phenomenon 'I exist because I doubt' was his line of thinking. To me it appears similar to the acceptance of a testimony by the judge only after it has stood the test of cross-examination by the opponent's lawyer. Even if a very good plan is put forward by a democratic government its goodness can be only better convincing if it stands after being subjected to all questioning by the opposition in the house pointing out all possible shortcomings.

It justifies the existence of the opposition. The Universe itself comes as a proposition before a philosopher and he puts it to the acid test of 'doubt'.

Descartes also thought that it was difficult to distinguish between the dreaming state and the wakeful state as far as perception is concerned. The dream is commonly called false because it does not tally with the wakeful state experience, but then since the two experiences don't resemble each other, why should only dream be considered false? It could as well be other way round. It is equally possible that what we experience in the waking moments may be unreal and our so-called real world may be a trick of our imagination and it may have no existence outside our mind.

The philosophy of Descartes is known to be clear recognition of scientific spirit. He is also known as pioneer of modern philosophy in the West. According to him the substance is that which does not depend on any other thing for its existence and extension [i.e. the capacity of having increased size] and is the fundamental attribute of a substance. In finding attributes of a substance one should take into consideration qualities which are obvious and unique. Thus the only quality which is unique in a substance or a body is that of extension. Sound, color, taste, smell and such other qualities can not belong to the body because they are not unique.

Empiricist thinker John Lock divided properties of an object in two categories, primary qualities that are size, shape, length, solidity, number and mobility. These qualities he thought were independent of the observer and qualities like color, sound, taste, smell etc are secondary qualities because they really do not exist in the object. I do not agree with this. Even what he termed as primary qualities depend upon our sense organs and to know shape, size, length, solidity, number and mobility we have to depend on sense of touch, sense of sight, for without these sense organs we would not be able to detect all these properties; Apparent size, shape and length depend upon arrangement of the refracting part of our eye. If our eye lens were cylindrical or of prismatic, the shape of an otherwise rectangular thing would be seen to be altogether different, people suffering from astigmatism see different shapes and sizes with respect to those not having this deformity. When tried to be felt by sense of touch, the shape and size would be perceived to be in the same shape shown by the eye having any of such curvature deformity, for the movement of the hand or fingers will be found to follow the lines and curves or edges of the object since it is the same eye which is judging the movement of hand or fingers. If an edge is straight for an observer ' A' and is curved for an observer 'B' due to different arrangement of the eye, the movement of hand shall also follow, while feeling the edge, in the same sense as the eye perceives the edge. If 'A' calls the edge to be straight and 'B' learns this term from 'A', 'B' shall also call it straight although the word straight would not mean the same actual perception of the object for 'A' and 'B'. In fact entire appearance of the world will be different if a sense organ is set in a different manner. Solidity is judged by the sense of touch and sensation of mobility is dependent on sense of sight and that of touch. If this were not true, the fact that Sun is stationary compared to the planets and that it is Earth which has motion, would have known right from days of Adam and Eve and world would not have to wait up to when Copernicus or Galileo declared this to the surprise of the people of the world. Thus all properties including what Lock calls primary do not really exist in the object. One can not validly argue that the present arrangement of our eyes, nose, or ear is the perfect one giving true picture of this external Universe. There are many animals that excel human beings in perception of external objects. Thus what we perceive is not what really exists. The qualities of the external objects, which our mind feels they have, do not really exist in that object.

Here question arises as to why a pen looks different from a bottle. If the observer supplies all the qualities they appear to possess then why pen should always appear to be as pen and a bottle as bottle? I shall answer this question in stages. The first stage reply is that there has to be something without perceived qualities but having power to excite the observer's senses and mind to make him throw upon it a particular set of qualities. The final reply which will be conclusive shall be given later after some points necessary to be understood before grasping are dealt with. Suffice it to say right now that if the substance were sum total of certain qualities imposed on it by the observer, the observer could throw any set of qualities of his choice on any actual or virtual thing situated in any direction he looks in. Thus one should be able to perceive a horse where a house is, but this is not so because something beyond qualities is there in the object which excites the observer to perceive the object with a particular set of qualities. That is why a tree should appear a tree when looked at it a second or a third time. It is wrong to think that the perceived qualities themselves excited the sense organs to begin the perception, for perceived qualities come into existence AFTER the perception has begun. The cause of triggering the perception of qualities has to go beyond the perceived qualities. Effect can not be or precede cause in any case. Bertrand Russell says that a substance is nothing but a bundle of qualities. A thing or a substance is taken to be the subject and properties form the predicate, but no substance can be imagined without any properties. Thus he believes that the name given to any substance is simply a name of bundle of qualities and nothing beyond it. But I totally differ from these views. The properties or qualities of a substance totally depend on the perceiving subject. They have no separate existence from what our sense organs and mind suppose them to be. The question is why should the substance begin being perceived by us? The qualities themselves can not initiate the perception because they are the RESULT of perception and not the CAUSE of it. Result can never be the cause of the same phenomenon. Thus we have to conclude that there has to be something beyond qualities which has the power to excite our sense organs and the mind and then cause a set of qualities to appear according to the capacities of the organs and the mind of the observer or the perceiving subject. The substance or a thing is therefore a combination of something beyond qualities and the apparent qualities as perceived.

However it has to be noted carefully that in dreams, when there is no perception of an external object, we create in our dreamland various objects of various qualities and it is only the mind which does it, and so long as the dream is on, these objects are as real as the objects of our waking stage which we find around us. During dreams all external sensing is closed and it is only observer's mind that creates the dreamland with various objects and it also connects all things seen, by a dreamland logic which is as convincing as our wakeful state logic.

Knowledge of an object is a different thing than from perception through our sense organs, although loosely speaking the two words may be used to convey the same meaning. If a camera is set with its lens focusing the traffic on the road, a long chain of images pass on the screen but camera itself does not develop any knowledge that men and cars are passing by. It is really the mind, which analyses or synthesizes the perceptional signals and creates what we call knowledge. When I see a car the signal of the image formed on retina passes to the brain through optical nerves and excites the memory cells of the brain. I compare the image of the car seen now with the cars seen before and the mind comes to the conclusion that the thing seen now is similar in general appearance and its other functions like motion or noise are also similar to the thing seen earlier and what is now seen is a car. Now this conclusion is really the knowledge that the thing seen now is a car. Thus it is the mental process which converts signals of the sense organs into knowledge. If we were not blessed with this in-built function of mind, we would not have got the knowledge of anything in this Universe. As far as converting perceptions into knowledge is concerned the mind may be called the sixth organ which generates knowledge of the thing perceived. Logically, if this in-built functioning of mind is differently set, our knowledge would also be altered or modified and in this regard the mind can be regarded as an additional organ like the well known five. And so this should go without saying that if any one or more of all these six organs were set in a different manner from what they are now set, our entire knowledge of the Universe will change. In dreams, when the mind is set to a different logistics, violation of the law of gravity is no surprise to the dreamer and he may think an apple going upwards as a natural phenomenon of his dreamland. Who knows our wakeful state logic is not defective? Unless we are able to know what is perfect, we are not right in making any comparison. However it is common practice to regard our wakeful state logic as the best and all philosophers and scientists proceed with this convenient assumption.

When mind has been seen to behave like other sense organs in the sense that inherent property of mind, to connect observed thing with memory and to convert percept into a concept, the conclusion that something exists in the object which is beyond qualities, but which excites the organs and the mind to perceive the object having a particular set of qualities, must also be classed as an additional quality perceived by the mind. Just like eyes make the mind detect the color, the mind makes itself detect the quality of the object to excite the observer to perceive the object. It is just a conclusion similar to judging color, taste or the sound of an object. Thus the quality of the object to support the sense-organ-perceived qualities is also now an additional quality perceived by the mind to be in the object. But examining this additional quality in further details we find that this quality was concluded to be present in the object due to the qualities perceived by the sense organs only. Mind has memory inbuilt which detects a horse at one moment to continue its existence in the next one. Thus mind as an additional organ concludes that there is something beyond the sense-organ-perceived qualities, which excites the observer to project a particular set of qualities on the object. For if the qualities in an object are entirely creation of the observer then he should be able to throw a changing set of qualities on an object thus changing a dog into a horse at his sweet will.

However like the five sense organs mind is also an organ with its own functions like memory, converting percept into concept and knowledge and the like. If this organ that is mind has also a different setting, conclusions are bound to differ. An untrained mind reaches different conclusions compared to a trained mind. Mirage is pool of water for an ignorant mind whereas a trained mind knows it to be result of refraction of light rays through layers of air with changing densities resulting due to variation of temperature. We cannot hold that even much trained minds are perfect minds knowing the entire truth for if they were, philosophers would not have reached different conclusions about the same problem and old scientific ideas would not have given way to new ones about the same phenomenon. Thus we cannot escape the conclusion that all qualities in an object are dependent on the observer only and are non existent in the thing observed and these qualities include even the additional quality perceived by the mind as discussed above.

Philosopher Berkeley (1685-1753) argued, while criticizing Locke, that there cannot be categories like primary and secondary qualities in an object. Color of an object can not be separated from its shape or form for the color takes the shape, which the object has. Whatever is true for color must also be true for its shape and size as well. Berkeley altogether denied existence of matter outside the mind. He maintained that material, which is non-mental, can not be the cause of sensation, which is mental, and as such presence of sensation can not be taken as proof of existence of an object. He evolved a system of subjective idealism and he believed that all objects to be subjective, to be dependent on mind. Thus sole existents in the Universe are ideas belonging to the minds. Kant (1724-1804) and Hegel (1770-1831) believed that it is the mind, which creates nature. However Hegel maintains that material of which nature is made does not come to the mind from external source but mind contains the substance with which nature is created.

I hold that idealism of Berkeley or Hegel touches extremes by denying the matter altogether and saying that only mind creates the Universe. If that be true then the question why one is not able to create things of his choice and demolish the unwanted by power of his mind, can not be answered. It is a fact that human being is also a matter and his brain is an integral part of him. What we call mind is nothing but property of brain to think and to have developed some concepts and some emotions mainly due to physical needs, as a combined result of which the direction of actions of the man are steered. If matter is to be denied then along with it go the brain and the mind too. Suppose an observer 'A' is watching an object 'X' and there is another observer 'B' who is watching both. Then for 'B' the object 'X' and the observer 'A' both are matter, the object 'X' having certain qualities imposed upon it and the man 'A' who is showing some other set of qualities including his qualities of being able to watch and describe how 'X' looks or smells or sounds like. Thus while denying matter, 'B' denies both pieces of matter 'X' and 'A' equally. If there is a picture of a house and a cow standing near it, then cow can not be more real than the house since both are part of the same picture. The observer 'A' can not be more real than the object. The chain of observer and the observed can be further extended by introducing observers C, D, E and so on till we exhaust all that can be classed as observers and we can not help thinking that an observer outside this Universe can only tell us whether there is anything real in this Universe or everything is nonexistent since the day it is supposed to exist. Some of the ancient philosophers of the East particularly those of India have thought on these lines but their convictions depended more on the intuition than on reasoning. I shall discuss this idea in details later.

Now coming to realism like that of John Locke and others which tells us that all objects are real having qualities perceived by the observer, and our sense organs and the mind are forced to register the objects with their qualities, is the other extreme end, leaving nothing for mind to have any creative role or the mind being left helplessly to behave like a camera or a recording machine. The best it can thus do is to analyze the observed properties and to find out 'How' or as far as possible 'Why' of such qualities. And that is what science has been and is doing and shall continue to do. It should not make any difference to the scientist whether the observer and the thing observed are both real and both false. Science shall reach a dead end only when one of them is real and the other false, but such a proposition has luckily not been proved beyond doubt so far, and thus exists science.

I hold that the truth lies somewhere between the extreme idealism and a similar realism. Although the objects don't really possess the observed qualities, there is something in them beyond all qualities, which excites our sense organs and the mind to begin throwing a set of qualities on the objects. This applies equally to all living and inanimate objects since an observer is an object of perception for another observer. The subject shall be expanded in later chapters.

I have included extensions of our sense organs in the meaning of sense organs as pointed out earlier. The advancement in physics was only possible due to such instruments, which extend our power of perception or the power of ascribing sense qualities to any object. When a distant stellar body is watched through the telescope, the effect is as if we were nearer to the body than we actually are and what we see is how the body would look if it were nearer. But here, perception through the telescope has the same limitations described in the preceding pages. Even if a radio telescope is used, it translates radio waves into perceptible form to our sense organs. A microscope can not add a new adjunct to the particle, which we cannot perceive through our sense organs. What applies to these elementary devices also applies to much complicated and sophisticated apparatuses used in research in particle physics or any other branch of science. Particles like electron which have a mass and a charge, and like photon which has neither, are detected through such fine instruments but finally the instrument brings the minute things in the purview of our capacity to perceive. These instruments do not create any additional sense organ beyond the five well-known ones.

I can not pass on to the next chapter without letting readers know how the existence of matter has been dealt by philosophers Spinoza, and the mathematician philosopher Leibniz of the West and in Nyaya philosophy of the East.

Spinoza (1632-1677) thought that there is no place for plurality or dualism in the Universe. In Spinoza's words 'If substance is that which needs nothing other than itself to exist, if God is the substance and everything else is dependent upon Him then, obviously there can be no substance outside God. Then thought and extension can not be attributed of separate substances, but are merged in God; they are attributes of one single independent cause and bearers of qualities and events, the one principle in which all things find there being'. Thus Spinoza condemns the dualism of Descartes. Descartes admits of two ultimate realities that are matter and the mind. Spinoza defines attribute as the substance as the intellect perceives it. He believes that God is without any attributes for attributes limit him, and only we see him as attributed. Spinoza's views somewhat look tallying with the Indian Vedanta to some extent, but the difference shall be clear to readers as they go through rest of the chapters.

Leibniz (1646-1716) regards the relation between body and the mind as previously established harmony created by God such that whatever happens to the body appeared to influence the mind and vice versa. Leibniz gives example of a watchmaker who makes two watches and adjusts them in such a way that they show identical time. Thus God created body and the mind in such a way that whatever happens to one influences the other. This is really very ingenious idea to explain apparent cause and effect phenomenon between body and the mind. In fact according to the example of adjusted watches, none is the cause of changes or influence over the other. It simply looks as if one affects the other, just as time shown by one watch can be misinterpreted to be the effect of the time shown in the other watch. About creation of the Universe and it's phenomenally I shall touch Leibniz's philosophy once more in chapters to come.

According to Nyaya philosophy, perception is unconditional knowledge, which arises out of the nearness or the contact of the object and the sense organs. Of course this does not include knowledge which is intuitive because this type of knowledge arises without contact between the sense organs and any object. Knowledge of pleasure or pain arises without any such contact. The philosophers go further deep and say that when the sense organ comes into contact with the object, it first of all only causes awareness of the object without any knowledge of its name or qualities and they call such awareness as indeterminate (Nirvikalp) perception. The existence of a particular object, say a horse, is proved not by perception but by inference. The indeterminate knowledge of the character of simple awareness soon changes into determinate knowledge when the observer becomes aware of the qualities of the object, and then he compares the impression with the past experience and develops determinate knowledge by inference of such a comparison.

Philosophers of this school include space, time and mind in the list of eternal things and say that these things do not need a material or efficient cause for their existence. Other objects, which are limited and not all pervading, must have a cause for their existence in their present shape. Without the guidance and directions from intelligence the things can not take the present shape in which they are.

The discussions in this chapter can be concluded briefly like this. What we perceive as objects of the external world is nothing but bundles of qualities through our sense organs. All scientific instruments are simply extensions or the enhancement of capacities of our sense organs. All complex qualities, all microscopic or even qualities of the elementary particle level are known through our sense organs operating through complicated scientific equipments or instruments. That these equipments or instruments do not add a new sense organ is as clear as writing on the wall. The objects are thus only bundles of qualities as grasped by our senses with the help of the mind (the sixth sense organ) as described above. However we do not know what the object really is apart from all qualities. The object by itself or in itself appears to be only a mystery so far.

How all the matter has come into existence is also not known. Even if it is assumed that energy has partly converted itself into matter, the cause of such a conversion is not known. If Leibniz's law of sufficient is applied it would lead us to the conclusion that matter, as it is really does not exist, for no sufficient reason to justify its existence seems to exist. Our minds conscious of our process and action of thinking, is not matter, but the brain without which mind is not supposed to function is definitely matter. So in case matter is non-existent we shall have to examine whether our minds could exist as they are or only a cosmic intelligence shall be operating in place of individual minds. Thus if any man is a matter to me he is known to me through his qualities like appearance, his speech, his actions and the like. If qualities are removed then the man and all other men with their minds vanish from my perceptions. Thus another question to be resolved is whether plurality of minds is real or there is only one mind. The entire world perception looks like a dream wherein we perceive only our minds, our thoughts and the entire dream land is like external objects in this Universe including trees, stones, mountains, rivers and oceans, other men and women, other animals, stars and planets and all. The man who dreams calls himself 'I' even in dreams, but this 'I' although born to the dreaming man, is not identical to him. The dreaming 'I' may be running away from a fearful dream tiger whereas the man who dreams is sleeping comfortably in his bed. The perceptions of the Universe are similar to those of dreams although the dreaming perceptions do not tally with those in the wakeful state in time, space and logic. Bernard Russell in his book 'ABC of Relativity' (1997 edition page 25) has remarked 'If there were no reality in the physical world, but only a number of dreams dreamt by different people, we should not expect to find any laws connecting the dreams of one person with the dreams of another. It is close connection between the perception of one person and the (roughly) simultaneous perception of another that makes us believe in the common external origin of the different related perceptions'. Here he tries to prove the real existence of the external world by stating that different persons are dreaming different dreams. He is oblivious of the fact that if this world perception is a dream, what he calls as another person is also a dream object just as a tree or a house for anyone's dream, and that the impression that different persons perceive approximately the same external world is also an impression in the dream world. He should not have separated the living things from the non-living things of the wakeful state or of the dreaming state. The external world consists of both. Thus it can not be proved by the argument that Russell puts up that the external world is real. It could as well be a dream, in which case the question whether there are many minds or only one mind stands resolved. I shall again touch the subject in chapters to come.

Chapter 2

The Time

First of all let us see whether there should be or might be or is a thing like time because Democrites thought that there is nothing like time, the reason being that the past can not be perceived and the future has never come and hence beyond perceptions. As soon as what we call future comes, it changes into unperceivable past. What we call present is only a moment infinitesimally small and hence not capable of measurement since it has no duration at all... We may be inclined to agree to this because only duration can be measured and for having a duration the beginning and the end are absolutely necessary and by the time we detect the end the beginning goes into the ever non-perceptible past and we have only left with the memory of the beginning, and memory cannot be one end of the time duration to help measurement, for a measurement the two ends must be of same nature and not heterogeneous things. We shall return to this point later in this chapter.

As put by a famous philosopher, 'if no one asks me I know what time is but if someone asks me to explain to him, I do not know'. Really most of us know in our minds what time is, without being capable of defining it. Earlier I had defined it as difference felt in mind between two events independent of the nature of events but soon I realized that the definition is more about its measurement than defining what time is. If events do not occur how would I be able to know the quantum of time?

Time is certainly not matter within the meaning of the scientific definition of the matter. For time to be classed as matter it should have weight and should be perceptible to our five outer sense organs which send signal of color, smell, taste, sound and of feeling of touch. Time does not occupy any finite space although it may look to be connected with the space. Time is not energy too since it has no capacity to do work although all work seems to happen in some duration of time. The physicists measure the work as mass multiplied by acceleration multiplied by distance and although acceleration contains time as one of the units this time unit is not the cause of acceleration. Acceleration is increase in velocity in unit time and thus after a body is accelerated the time comes into make it capable of measurement.

Time is not perceived by our outer sense organs but we perceive it through the sixth inner organ, which is mind as discussed in chapter 1. We perceive time by our change in mental and physical state that occurs between two perceptible events. We perceive milk pot put on the stove and we perceive its boiling. Between these events our heart beats say three hundred times, our blood runs miles, our nerves send many signals to our brain, many thoughts come to our mind and go and we are conscious of the changing state of mind due to the power gifted to us called memory. This consciousness about our changing state of mind gives us the idea of something that elapsed between the two events and we call that something as time. This ever-changing state of mind goes to the memory and gives us the idea of what occurred earlier and what later and since we have to use this eventfulness in life while interacting with others in our day to day work the word and the idea of time gets introduced. Thus as a preliminary answer to the question in the beginning of this chapter it can be said that succession of events have to be expressed in terms of time as explained above although a thing like time might not exist by itself in separation from events.

I have got a copy of Oxford Concise Dictionary, a 1994 print and therein the word time is described to mean, 'The indefinite continued progress of existence, events etc in past, present and future regarded as a whole'. This meaning of the word time is neither fully wrong nor fully correct. Time is shown to be inseparably connected to existence and events which is correct, but the use of words past, present and future should not have been included in defining time, since these words depend on time itself. Thus time should not be defined in terms of something dependent on time itself.

The question is that if events don't occur will time really exist? While we watch a photograph wherein no events occur we still perceive the time for which we stared at the photograph. If events don't occur in this Universe the entire perceivable Universe would be like a still photograph. Will the time stop then? One would like to say that time will still continue to pass with changeless world but while saying this he will unknowingly assume that observer is something beyond the standstill world just as those who watch the photograph are not a part of it. He assumes that his own mind shall continue to work, his heart shall continue to throb and he continues digesting his food while he is observing the standstill Universe. He shall estimate passage of time by means of fatigue of his brain due to such watching of that large photograph of the world or by his heartbeats or by the hunger he feels after he has digested his food. But man is as much a part of the world as the blowing air or the growing tree is and as such with a standstill world he also becomes standstill, without heartbeats or thoughts or any fatigue and shall never be able to notice, measure, or define time, since he will not be able to think at all about anything. His thinking comes to halt because thinking itself is series of events in the mental status. Stop thinking and there is no time. Begin thinking and time comes into existence at once.

Another question is that when human race is of recent origin on this globe, how we can say that Earth was formed so many million years ago or that Sun was born so many billions years ago? Perception of time through our organ known as mind is an elementary phenomenon as every body knows. There was no observer when the stars were in their making or the Earth was taking its shape. John Wheeler says, "No elementary phenomenon is a phenomenon until it is an observed phenomenon". However scientists make an estimate of time that must have elapsed since the birth of the Sun or the Earth. It is stated that Earth was born some six million years ago, without there being any one to observe. Such statement indicates that time was there before human race came into existence and how it should co-relate with the fact that perception of time and hence existence of time depends on human thinking, needs to be examined. The answer to the question as to how the age of the Earth or the Sun could be assessed is not far to seek if we believe that logical conclusions assuming all affecting parameters are known, tell us the truth. Physicists observe certain changes in the positions of stars, planets, comets or even galaxies, they observe the radiation's emanating from these stellar bodies, and they observe all such things which to the best of their present knowledge are related to the indications about evolution of the stars or planets and then by use of mathematics they extrapolate the observed uniform changes backwards and tell us that such an event say birth of the Sun must have occurred so many millions of years ago. Thus the estimate of the quantum of time that might have passed between such astronomical event of the past and the present day is based on the logical thinking of the scientists today and clearly enough, an estimate is not at all a perception of the time of such event of the remote past. The estimation of the time of events of the remote past is dependent on the assumption that the nature works uniformly and mathematical formulae based on this observed uniformity at present and the supposed same uniformity in the past, give us the truth, It is assumed that human thinking at present could dictate the happenings of the present and future and that it leads to correct estimate of the happenings of the past so remote. Suppose this uniformity was not there in the remote past, our conclusions based on such uniformity observed at present are bound to be wrong. These assumptions may not have been explicitly articulated while working out each step in such a computation, but human mind has got an inherent tendency to be satisfied if something is found to be uniform at present and to unconsciously assume that there should be nothing to stop it to be so in past and future too. No one has observed when Sun was being born out of stellar dust due to gravitational force which brought particles closer so fast as to raise the temperature to the level fit for fusion of hydrogen atoms emanating huge energy. This present day theory of formation of stars is another example of extrapolating backwards the natural phenomenon observed today. Had we not had the power of thinking and laying rules of logic, there would neither be the knowledge of today's perceptions nor an estimation of the past. Both evaporate if thinking stops. As already said, it is to be marked that in extrapolating things, we assume that nature is uniform and the reasons that existed to create a past event if continue to exist in future, the same or exactly similar event should occur again. This assumption about uniformity of nature may not be flawless; for even we want to test it for future but it is impossible to do so. We can only test it for the past events only. Whatever has occurred in the past does not by itself prove that the same occurrence shall be in future too. That Sun rose yesterday and today, is not itself a proof that it will rise tomorrow. No phenomenon can be a proof of its future occurrence. But scientists presume principle of uniformity of nature for it has not yet grossly deceived them in the past. But however if the world becomes eventless there is no thinking and there is no time.

Thus time has to depend on thinking minds and nothing else in this eventful Universe. The question again is raised whether the time causes events or events cause time. In the West two basic theories were put forward in the ancient past and since then the philosophers of the West have been racking their brains on criticizing them and attempting modifications that would look more plausible than the original theories themselves. Simply stated the two theories are as under.

(1) Aristotle's principle which says that there is no such period of time as not to have any changes, and upon this is framed the argument that when we talk about any temporal item we have to refer to some occurrence or event at that temporal item. A particular moment can be identified by referring to some other item or happening at that moment. For example when we refer to something happening at 3.00 P.M. we mean that two things occurred at that moment, the first is the moment of the happening and the other is the clock striking or showing 3.00 of the afternoon at that moment. In another way this argument implies that there should be no perception or even existence of time without a change in the worldly things. Changes include change in the mental status of the perceiver too.

(2) The other theory is that time is something flowing from even before the world came into existence and shall continue to flow even

if there is no world or in other words the time is independent of events that occur in time span. This view is commonly known as Platonism.

In this reference however the term event and duration shall have to be understood. Point of time is like a geometrical point in a line, having no dimensions at all. There can be no event at any point of time because for an event to occur there has to be some change in situation or properties or magnitudes of things. For a change at least two points of time are necessary, the one when the event begins and the other where it ends. A number of points of time when added together can not make duration because addition of millions of zeros does not make a finite quantity. Thus some philosophers think that if time is a result of joining moments together then a moment too should have some definite, howsoever small, duration. Some are inclined to think that there can not be a smallest unit of time less than the time taken by light rays to cross inter-atomic space.

However I hold that time and space are continuous entities and any attempt to show them as summation of points of time or points in space is futile. The motion of a body since it depends on these two continuous entities is also a continuous entity not capable of being shown as summation of different position the body takes at different moments or points of time if that phrase can be used validly. On the other hand the matter is not continuous but is discrete. A body of matter can be shown to be summation of elementary particles, which constitute the atom. The point is more clearly explained in chapter 1

I have purposely avoided special philosophical terms used in rigorous definitions of things above so as to let readers save their energies to follow what is coming hereafter. Scholars of philosophy may find the above to be a far simplification of Reductionism and Platonism.

Anyway, the question in a nutshell is whether time causes events or events cause time and it still looks to be unsolved. Passage of time and measurement of time can be separate things if both were not connected to the same observing mind. If time were not at all passing for some duration, how would we know of it? Suppose our clocks show 12.00 Noon and for a time which would occupy five minutes, the time stops. Then neither hands of the clock would move nor our minds would work for that duration and thus this interval shall pass unnoticed and we would perceive hands of the clock continuously and uniformly moving as they were doing earlier. In fact we measure and perceive time only through changes.

Take the simple question, when did the time begin? Imagine one million years before Christ or a trillion years before that so long as we use the word 'before' it is our imagination which we stretch to go much behind say Christ or Moses or even Adam and Eve. In Genesis the origin of firmament, Earth, light etc has been attributed to Gods wish. God said 'Let there be firmament' and firmament was there. However there is no mention how the time began. And when God said something time must be already there so that something could be said. Saying something needs uttering words after words and even thinking to say something, needs ideas after ideas, and this can not take place without time already existing. Thus if time really has a beginning, when this should have been, is not clear. Some ancient Eastern philosophers think that just as we can not know during the dream as to when the dream began, we in wakeful state can not know when not only time but also the entire panorama of this Universe came into existence. Anyone can circumvent the view that when God said something, time must have been already there by arguing that God does not need time to think or say something. He is omnipotent. When he said firmament should come into existence, time too must have begun at that point. Today's Physicists too say that time and space are so inseparably connected that the two so far independent entities are now fused into one called space-time. Thus it could be said that space-time, Earth, waters, day and nights have come into existence as and when so imagined by God. I shall go into depths in chapters to come as to how and in what case imagination could take the shape of a material Universe with large matter bodies and energies so apparently systematically interspersed in space-time.

The question whether time had a beginning and has an end has been tackled by many thinkers. If the time is beginning-less and end-less it can be represented by a straight line extending both ways to infinity. Although it may not be justifiable to think that time is represent-able by geometrical figures, still some thinkers are prone to try to represent any entity whether material or abstract by such figures. Moreover it is a style these days to think of a circle or a sphere when endless things are to be shown to be finite in expanse. Circular closed time is another representation of endless time. However it has a defect of its own repetition over and over again as we move along its circumference and it becomes a problem as to whether if segment of the circle is repeated, the events should also repeat. If time requires for its existence the events then events should repeat, but if time is something independent of events then although time as an entity may repeat without past events contained in it.

Whether time is to be identified with events or not is yet settled question among Western thinkers. Some favor Reductionism and some Platonism but number of thinkers choose a noncommittal status by saying that either of these two does not tell the truth. But then what the truth must be is also not explicitly and unequivocally articulated by such thinkers. There is an increasing inclination in both physicists and philosophers to now regard the perceiving mind to be the center of the question and in that they can be considered to be coming closer to the Eastern philosophy wherein the mind is considered to be the cause of so called creation. William Newton Smith concludes his famous work 'The structure of Time' (published by Routledge and Kegon Pauls, London) with the remark 'And even if the views advanced in this work do make some aspects of time less mysterious, its promiscuous character means that there are depths yet to be plumbed, in particular the perhaps most puzzling aspect of time, the relation between time and the consciousness remains'. Even physicists are gradually becoming conscious of the role of human consciousness in perceiving and developing concepts about natural phenomena. All what is called knowledge depends upon the perceptible qualities of matter whether living or inanimate, and the realm of any actions or perceptions of matter whose properties we are unable to perceive through our sense organs including the mental faculty of logic is permanently hidden from a physicist or a layman's eye.. If matter has got some properties which we can not perceive by our sense organs or their extensions then we are not able to apply logic too to form any concept of them. Whether a supernatural being who could see the Universe without being a part of it, can exactly develop a unified theory covering all aspects of perceivable Universe is still a matter of speculation particularly for Western thinkers. Readers will find this aspect of matter having properties beyond perceivable ones, is dealt with in details in chapter 1 and I shall again return to philosophical aspects of time later in this chapter

Now let us see what physicists have to say about time, because it is they who use time element in almost all formulae in physics. Velocity, acceleration, momentum, force, work done, energy, and power are their fundamental entities upon which are based almost all their equations about moving globes in this Universe. Even calories, which represent amount of heat, are convertible into its equivalent physical unit of energy, and so also the light radiations or sound waves. From ancient past to about 200 years ago, philosophy and science or mathematics was not much different from one another and each one seemed to either supplement or include the rest. The Greek philosopher Aristotle gave arguments to prove that the Earth was not flat but is spherical. He watched the shadow of the Earth on the Moon. This was an astronomical observation, which a philosopher made as a part of philosophy. After Aristotle the philosopher Ptolemy in the second century said that Earth was the center of the solar system and the planets and the Sun went round it in spherical orbits. The succession of musical notes in an octave is also attributed to Ptolemy's talents. In the recent past philosophers like Descartes, Laplace and Leibniz had contributed much to mathematics and physics. However during these years the science became more time occupying due to various discoveries with the result that philosophy and science or mathematics began looking distinctly different from each other and now to such an extent as to make some scientists and some philosophers as opponents of each other. I do not hesitate to call a scientist a philosopher and vice-versa. Philosophy is the treasure of knowledge to which both add immensely, piously and selflessly. Galileo, Kepler, Copernicus and Newton are more commonly known to be physicists or astronomers than as philosophers although they immensely contributed to search of ultimate reality of nature which forms major ingredient of the definition of philosophy.

Physics and astronomy has more systematically developed in the West than in the East. It may not be of relevance now to go into how these sciences developed in the East say before 400 B C, since now much can be accurately predicted by modern laws of physics and astronomy than by those of the ancient past. Aristotle was the one who affirmed that Earth was spherical and then it was said by Ptolemy that planets orbited round the Earth in spherical orbits. Nothing notable seems to have been added in a period of about 1300 years until Copernicus, Galileo and Kepler came with new model of what we know as the solar system now when they said Sun was the center around which the planets including Earth went round in regular predictable paths or orbits. Then it was Newton who gave the world equations, which described mathematically Laws of motion and Laws of gravitation. It was a great landmark because Newton's laws accurately described how and why objects should move and why planets should go round the Sun without falling into it or without flying away from it. In all such equations governing the motion time was supposed to be absolute and the same whoever measured it wherever. It was considered independent of space and matter. It was taken to be independent of the velocities of bodies on which it is measured. It looked as if the Universe was a big machine, and all motions of bodies in it were by themselves obeying Newton's Laws and had little connection with the so far unfathomable mysteries of nature attributed to the powers of Almighty. The following lines depict how poets took these scientific advancements,

> And of old from Sinai top
> God said that God is one
> By Science strict so speaks He now
> To tell us, there is none

Earth goes by chemic forces
Heavens, A Mechanic Celest!
And heart and mind of human kind
A watch-work as the rest.

And to top this as if, was Alexander Pope's exalted eulogy epitomizes Newton's impact on his age,

Nature and Nature's Laws lay hid in Night
God said, let Newton be! And All was Light

Due to Newton's laws of motion, the ideas of momentum, force, work and energy could be made capable of being quantified, and his law of Gravitation explained the motions of orbiting planets around the stars like our Sun. This was a great contribution to physics and it was felt that nothing new could then be known, since these laws almost solved all riddles of motions known at the time. Newton was however aware of the fact that it was not very rational to think that force of gravitation could be experienced instantaneously by bodies at large distances from one another, but his calculations about motions of planets by his own equations were tested to be true. In all Newtonian physics, the time was taken to be absolute entity, and did not have any effect on it whatsoever due to motions of bodies in space. The time required for bodies to reach from one place to the other, and all observed phenomena were independent of the observer wherever he was and whatever motion he himself had relative to the bodies in question.

Albert Einstein shook this idea of absolute time in the beginning of the twentieth century and from his time up to this day of physicists John Wheeler and Stephen Hawking and others; physics has put up many amazing things before the world to ponder upon. Einstein said that if a body moves with a velocity comparable to that of light (186000 miles per second or 300000 KM per second) the time for this moving body dilates compared to the stationary body. Further the mass of the moving body also increases till theoretically it would increase to infinity if the velocity of the body really equals that of light. In that case the unit of time say second shall dilate limitlessly and the flow of time shall cease to be felt. However all this happens to the moving body only if observed from the stationary body in relation to which the motion is, while on a stationary body one goes on counting seconds and the stationary body continues to have a finite mass. What led Einstein to formulate his special theory of relativity was the amazing fact that velocity of light did not change even if the source of light or the observer move at a considerable speed relative to each other. In fact if the source of light is receding fast from the observer the velocity of light should measure less than the standard value in stationary conditions and if the source were approaching the observer it should measure more than the standard value. But experiments conducted by Michelson and Morley took the physicists by surprise when they found the velocity of light was the same irrespective of the relative motion of the observer or the source. To explain this Einstein thought of a change in the other ingredient of velocity, that is, time. He came to the conclusion that if a body A is moving with velocity comparable to that of light with respect to B the dilation of time would be given by the relation $t/T = 1/\sqrt{(1 - V^2/C^2)}$, where V is the velocity of A and C is the velocity of light and t is the time at A and T the time as measured at B. Thus the Newtonian absoluteness of the time was shattered.

In his General Theory of Relativity Einstein said that gravitation was not a force of attraction between two bodies but the effect is the result of warping of the space-time due to presence of masses and energies in the space. Planet which looks orbiting round the Sun instead of flying away from or falling into it is not due to force of attraction or the balancing centripetal force as supposed by Newton, but it was due to a motion in a straight line path in the warped four dimensional space time which gives an appearance of around in three dimensional space. Stephen Hawking explains this nicely by stating that that the phenomenon is like moving shadow of an aircraft flying in a straight-line path in three-dimensional space. When the shadow passes over the undulating two-dimensional Earth surface, it seems to move on a curved path.

To have an idea how science thinks about time, we shall have a very brief description about how a star is formed and how it dies. Formation of a star begins as large amounts of hydrogen particles or molecules start falling on one another due to force of gravitation. As the gravitational pull goes on increasing the particles star collapsing with greater and greater velocities thus giving rise to huge amount of heat. As the big lump of colliding particles heats up to a very high temperature, nuclear fusion begins and hydrogen atoms form new molecules called as those of Helium which itself is an inert gas. This formation of helium from hydrogen release very huge amounts of energy as would be released from exploding hydrogen bomb and this energy is the cause of all light and heat radiating from a star. Our Sun has formed likewise and still it has enough of hydrogen as its fuel to emit light and heat for many million years to come.

In the life of a star a stage comes when all of its fuel is consumed and the star starts cooling. Due to fall of temperature the pressure of the gases which had kept the star in a stable condition against collapsing due to gravitational force, the star begins to contract. The limit to which it should contract depends upon the mass of the star. As described in chapter 1 the atom is now known to consist of many elementary particles and the well-known three, that is, electron protons and neutrons as thought earlier. Scientist Pauli discovered that two similar particles could not remain at one place at any time. Thus when star contracts the similar particles coming closer start exerting a repulsive force on one another thus counteracting the gravitational force causing the contraction. An Indian scientist Chandrashekhar calculated the maximum limit of the mass of the star, which would collapse into a stable small star, called 'white dwarf'. In such a case the repulsive force between particles helps maintaining a stable state. However if the mass of the star exceeds the Chandrashekhar limit, the repulsion between particles can no longer fight back the excessive gravitational force and the star collapses to infinitely small size exerting tremendous gravitational force. This extraordinary gravitational force does not allow anything including even light to escape the star and thus the star becomes a 'black hole' as named by the physicists. No mass and no radiation can go out of its field and whatever comes within its action area is drawn into it and vanishes, only to increase the mass of the black hole.

The process by which a black hole is formed as briefly described above is the result of research work done by Einstein (1915) Chandrashekhar (1928), Sir Eddington (after1928) Oppenheimer (1939) Landau, John Wheeler, and recently by Roger Penrose and Stephen Hawking (1965 and onwards). It is to be understood that actual collapse of a star to form a black hole has not been observed accurately, and whatever condition that might be inside a black hole is a logical guesswork based on present day scientific theories and mathematics. Tremendous gravitational pull inside the black hole should destroy or annihilate any mass that got trapped into it. The black hole is a point mass with infinitely large density and infinitely warped space time according to Einstein's theory of relativity.

One may think why all this about formation of black hole should be necessary, but as stated earlier, it is here that time comes into picture; physicists say that if a man is caught in a black hole, there shall be the end of time for him. However time shall continue to be for all other stars or planets outside the black hole. It is also said that some day if entire Universe becomes a black hole by engulfing all stars, planets, galaxies, black holes and the like, then time for this Universe shall come to an end. It is said about the black hole that that it emits radiations originating from its periphery known as 'event horizon' and the size of the periphery or the size of the black hole, that is, volume contained in its periphery depends upon its mass and if more matter falls in a black hole the boundaries expand and its mass increases. The mass converts into energy and that may be the cause of radiations according to scientists. In this Universe elementary particles and their anti particles are always moving here and there. They meet and annihilate each other causing release of energy and the reverse process is also going on in the space. If one of the particles out of the pair is caught in a black hole the other one being diverted outside the black hole must be causing such radiation from the 'event horizon', it is said. Any man caught in a black hole his mass converting into energy, is 'recycled', a term used by the noted scientist Stephen Hawking, and thus such a man exists in the form of matter or energy. It is also guessed that if a black hole has a decreasing mass then it may end in a big explosion and 'evaporates' that disappears completely. This may take many thousand billions of years or some times less than that.

When time, as a meaningful term is under consideration the so-called end of time for someone caught in a black hole is without any consequence. The time continues to 'flow' right from formation of a star to formation of a black hole and continues till its disappearance. If scientists can calculate this time for a star then even if the entire Universe collapses as a black hole, the time taken for complete recycling of its mass to form a probable new Universe can be calculated too. Thus time is much larger entity to contain all such changes and the idea that time comes to an end in a black hole or at the end of this collapse and recycling can not be digested. If this Universe shall be thus recycled over and over again forever then also time shall continue to be to witness all such tremendously thrilling events if at all they occur. And who knows there aren't Universes outside ours, which have a flow of time even if this Universe of ours comes to a singularity. Scientists and philosophers have never denied the possibility of existence of other or many Universes beyond our own of which we do not know even origin and the end precisely (and still call it ours). In one of the old Indian work 'Yoga-Vasishtha' the accomplished teacher of Rama describes how King Lavana spent years together as a husband of a tribal woman, and got three children and finally saw a dreadful sight which woke him up from the brief nap lasting for moments only while he was sitting in his royal court. The story indicates subjective-ness of time, and also dilation of time, when few moments were felt like passing many years... At least there was an idea about dilation or contraction of time may it have its roots in intuition rather than reasoning, and intuition has not been discarded even by modern scientists. Anyway we agree that time may dilate depending on the velocities of bodies on which it is measured and is compared with the stationary clock. But this dilation is in comparison to some other stationary base. Further if it were said that time comes to an end inside a black hole but continues outside it, then again this end of time is in comparison to some other continued status and is not absolute.

Looking to all above recent development in science and scientific guess work (say about ultimate fate of a black hole) the remark of Bertrand Russell, that 'Physics of Newton considered as a deductive system had a perfection which is absent from the physics of the present day' given some seventy years ago, still seems to be meaningful. Not only this but also such remarks shall go on repeating in future, so long as physics goes on developing.

The question here is that if such remarks go on to be justified forever then there is no end to the scientific knowledge at all and man should think of abandoning this pursuit of this mirage-like thing. The world was simpler in Newton's age than what it is now, what the complications rendered by computers, cellular phones and fast air crafts in everyday life today. So man can very well choose stop going further in physics and limit himself to only pragmatics like medical science or applied chemistry and the like. But this will not happen. Something imbibed in human nature, may it be called God's will, pushes the man ahead in the quest of knowledge of pure sciences like physics, mathematics or astronomy.

It is believed that space-time began when the Big Bang occurred. The singularity turned into plurality by some mysterious force causing the infinitely dense matter to explode; this theory looks to have supposed pre-existence of space other wise where would the broken pieces fly? Physicists do have an answer to this question. They say that since space came into existence with the big bang, there can not be any place where the point mass was situated prior to the big bang. It could be supposed to be everywhere in the now recognized space. But there are other questions difficult to answer as yet. Why should the point mass explode at all? What is the cause of the immense force of which we see remnants in the form of receding galaxies and orbiting planets leave aside the energy loss due to other effects? If the singularity behaved like a black hole then time after the big crunch up to the Big-Bang must contain the events that occurred between these two ends. It cannot be denied that space, time and matter and the causation that is chain of cause and effect are four things fundamental to the existence of the Universe. Wherefrom and why the infinite mass come into existence is also as unknown as the origin of space and the energy to cause matter to blast. Thus the Big-Bang theory does nothing beyond trying to explain present day configuration of stellar masses and their motions. It has not got an inch even, to explain origin of Universe. If the origin of mass space time energy and causation is to be attributed to God's will, then the scientists must choose to forget their few centuries old difference with philosophers whose theories being based more on mental exercise than on observations for which they are criticized by scientists. At least at present science has not crossed into the mysterious realm often taken to be God's equally by scientists like Stephen Hawking and philosophers like Leibniz, Hegel and Kant of the West and the philosophers of Upanishads in the East. God's domain looks like horizon. When you get near there it appears to be still farther.

It can be gathered from above that modern physics although tells us lot about elementary particles which constitute atoms and informs us about many deeper aspects of energy and mass and has explored deeply into astronomical phenomena, it does not tell us anything about beginning or end of time. Whatever feeble attempts have been made to connect everything to the Big-Bang theory they are without any consequence since they are not able to tell anything about time taken as a meaningful entity. If there are other Universes like other stars outside the black hole then just as end of time in a black hole has no effects on the other star which continues to live in time, the end of this Universe into a singularity is of no effect on the time that continues in the other Universe. Further the explosion of black hole into complete disappearance in a calculable time is another failure to find out real beginning of time.

Let me now return to some philosophical aspects of time. As already said earlier the questions like beginning of time or space or the chain of cause and effect commonly known as causation, had been subject matter of thinkers of the past and are still being attempted to be resolved by philosophers and scientists of the present. In the West St Augustine seemed to solve this riddle by saying that there could neither be time or space before the creation of the world, and therefore it is irrational to think that world was created in time. The world had a beginning but not in time, for time was created simultaneously with the world. He also said that time is memory. Remembrance is the past and expectation is future. Physicists might have, it seems, taken a tip from this Philosopher of the ancient past or equally well it could be his own intuition when he says that the Big-Bang must be the origin of Universe as well as time and space. St Thomas Aquinas says that there are certain truths for which we must depend on revelations, for they are beyond logic although not contrary to the reason. According to him God created the world out of nothing. Physicists of today however think that a point mass with infinite density must be the material of which the Universe is made.

Indian Philosophers of the ancient past have unequivocally said that time, space, causation and matter are the fundamentals of the Universe and the apparently perceptible Universe came into existence along with these ingredients. The philosophy of Upanishads however seems to have originated from intuitions of Rishis (the enlightened learned ones). It must however be understood that intuition must have been taken as last resort when all feats of reasoning and brain scratching arguments had failed to reach the solution of the problem of origin of the time, space, causation and matter. If reasoning was able to reason out its own origin then philosophers who claim that reason to be the only method of knowing things would not have differed so much in the final solution of the problem. But still the 'reasoners' may satisfy themselves when reasoning leads to the conclusion that reasoning alone can not solve this problem of origin of Universe and in that sense the necessity of taking resort to intuition is logically established. I am of the opinion that St Augustine or St Aquinas must have reasoned hard, before they came to the conclusion that certain things are beyond reason and only revelation or intuition can make them disclosed to the mankind.

Whatever we know about time is through certain qualities or adjuncts attached inseparably to it and as in case of matter the real thing beyond all qualities is still mysteriously hidden from us. We find events happening in time. There is no event, which could not be connected to some other event indicative of time, say, the Sun rising at 6 o'clock. We have memories of events witnessed and we can express the before-ness or after-ness of such events in relation to some other event. We can predict at what time a car should reach its destination if we know the speed of the car and the distance it has to travel. All predictions of future events based on scientific truths and derived by mathematics are found to be true when its time of happening comes. Whatever is known to us about matter, space or time is only through its properties or qualities which are perceptible to us. Time as an entity may not be directly perceptible through our sense organs or the mind, but events, which occur in time, are perceptible to us. We can remember the past ones and predict the future ones if all the parameters governing the happening of such events are known to us. I put it in this way: Time has got the quality of having events occurring in it and it is through this perceptible quality that we come to know or infer existence of time as a fundamental entity vital to the existence of Universe. However it is certain that without events being perceptible we can not know what time is, just as disregarding all perceptible qualities of a material object, we do not know what it really is. Existence of space too is known to us through its various qualities, say of containing matter in it or permitting motions of material bodies within it.

Returning to the question whether time causes events or events cause time, I hold that the former is true. Philosophical thinking must not be allowed to go ahead without having a peep into the scientific discoveries, for such discoveries reinforce the way to our goal of finding the truth. Even a physical fact, wrongly understood at one time and correctly interpreted much later, adds valuably to the treasure of knowledge in its both phases, first by invoking objections or criticism and later by appeasing the inquisitiveness. Newton's law of gravitation and Einstein's general theory of relativity have both added to the sharpness of reasoning and removing the doubts in last two and quarter centuries. Neither is less important than the other. The special theory of relativity tells us that time dilates for a body moving with speed comparable to that of light, relative to an observer on a stationary body. If it were possible to observe the events or happenings in the moving body from the stationary point, it would seem that events are occurring at much slower speed. The time itself dilates and so the events are displaced farther in time, and appear to happen slowly. The famous Twin Paradox which is used to explain the time dilation is that if one of the twin stays on Earth and the other goes on a trip in a spacecraft moving with velocities comparable to that of light, the time dilates for the moving brother when observed from Earth, and the touring twin would return to Earth younger than the brother who stayed back. The traveling man would spend less number of years compared to the proper time on Earth and this dilated time keeps him younger than his partner on the Earth. It indicates that as time dilates the physiological and metabolical functions of the body slow down compared to the normal conditions at the stationary point. Since the time itself dilates this slowing down of bodily functions would pass unnoticed to the person having this change in him. It proves that the time causes events. Further, take example of heat and its measurements. Heat causes temperature of water to rise and that is a necessary quality of heat. The quantity of heat is measured by observing the rise in temperature of a definite quantity of water. One calorie of heat is equal to what is necessary to raise the temperature of one gram of water through one degree centigrade. To measure any quantifiable entity, the necessary effect, which is caused by such entity, should naturally be the basis of its measurement. If rise in temperature were not the necessary effect of heat, the former could not have been used to measure the latter. The time is measured by events like moving of hour hand, minute's hand and the second's hand of a clock. A thing is measurable only through its necessary effects. If events were not the effect of time it would not have been possible to measure the time through events. If events were not caused by time and could occur independent of it, there was nothing to stop the physiological functions of the body of the traveling twin to happen just like those of the brother who stayed back on the Earth. They could be still going on in the same manner and at the same speed as on Earth. The events in the spacecraft like aging of the human body were slowed down due to dilation of time only.

A subatomic particle known as muon is also formed in the atmosphere due to action of cosmic radiations on the air molecules, and these muons travel very fast with velocities about 99% of the velocity of light. The muons produced in the laboratory however have much less velocities. Their life too is very short. It has been observed that muons traveling in the atmosphere with very high velocities seem to live longer than those produced in the laboratory with much less velocity. This is the experimental verification of the theory of time dilation. If there were no time dilation the muons rushing through atmosphere towards the surface of the Earth would have been seen with the same short life span as for those produced in the laboratories. If events could occur the way they like without regard to time there was nothing to prevent the faster muons to have life span equal to those with slower speeds without showing delayed decay into another type of particles. It is to be noticed here that special theory of relativity regards the space and time as one entity called space-time continuum. The time observed to have passed between two events is measured to be different by two observers moving with different relative velocities from the place of events in space. This does not in anyway affect the principle that the events are caused by time. The time may be different for the two observers, even in the scales they measure the times with, the events do occur due to time perceived by different observers. When a body moves exactly with the velocity of light the time dilates infinitely and the flow of time as if stops. When there is no time flowing the events too stop for the observer who observes the moving bodies with velocities equal to that of light, although it is not possible to actually achieve this velocity due to mass approaching infinity, requiring infinite force to push the bodies to these ultimate conditions.

A corollary of the principle that time causes events to occur would be that all the events of the past and future are just already there. They do not develop as one would normally think. This also goes to confirm what I have said in the chapter of 'Cause and Effect'. The Universe is a succession of disconnected events, which only human minds try to bind them by chain of what we call cause and effect and this tendency of human mind has its birth in repeatedly observed pair of events in succession. The space time continuum is four-dimensional. The time is regarded as the fourth dimension for convenience of understanding the effects of this four dimensional behavior of the Universe. The human beings can see or perceive three dimensions of the space and only one point of the fourth dimension, which we call 'now'. We are unable to perceive the past or the future and can see only the immediate present. The past and future of our everyday talk is simply memory and the possibilities or probabilities, and more than often we find the probabilities and the likelihoods deceive us. If we would have been clairvoyant enough to read the future we should turn either a prophet or a mad man having lost all interests in the life due to no charm left with future's uncertainties. Not only we but also even physicists have not been able to peep into the future with all their mystery devouring mathematics and stunningly complicated scientific instruments. So no reason to be dejected on that score.

How we are not able to read the future can be better explained in this way. Imagine that we are able to move in or perceive only two dimensions of the space, like imaginary flat creatures living on a tabletop. Suppose that a film of time having all events painted on it in succession is moved in the third or the vertical dimension gradually and vertically downwards at the edge of the table. We would in that case see only that painting which is in the plane of the tabletop for we are deprived of the capacity to look up or down which is the third dimension. The painting or the event which comes momentarily at the edge of the table is the 'present' for us, and the film above the edge is the future and the part of the film downwards is the past. Suppose all of a sudden one of us gets the power to look in the third dimension that is the time dimension, then all the future events become directly known to him while the rest of us only shall wonder how astrology can help us to grasp the future. Suppose further that the enlightened one gets also the power to move in the vertical direction, then he is the only authority who can directly tell us whether the two dimensional world of ours is limited in or is without beginning or the end.

Well, this opens up a new subject whether all events that occur are predetermined, and if so we have to do nothing from our own initiative to improve the world, which probably all our political leaders and governments take to be their privilege. We can be carefree about our own and our children's future, for what is to happen shall happen despite whatever we may plan. It is however difficult to experimentally verify if pre-determinism is true or not. We shall have to know in advance what is predetermined to happen, and then we have to wait to reach to that point of time when it is to happen and for this we must know the point of time when the predetermined thing is to occur. All these data are impossible to be available. According to pre-determinism whatever we do now is also predetermined, and if so then why punish an offender and why reward a meritorious deed is impossible to rationally explain. Indian philosophers of the remote past have propounded another theory for this. It is said that we are free to do any act on our own, but whatever we do shall have its effect in future, and one has to bear in mind this effect when he does any act. It is also said that the effect may not occur in this life of ours and it may have to be suffered or enjoyed in one of our re-births to come. If a good deed goes unrewarded for the entire life of a man, or a criminal is found to escape a punishment up to the end of his life, we may be inclined to believe in this theory of effect in the coming births, but this theory can not be verified for similar reasons. There is no definite formula as to, which acts gives what results, and then it is not known when the effect is going to be seen or felt. It is also not known whether the effect shall occur in this life or some future life, and then who knows one will be born in what form, and how shall he be identified in the next birth to be the same man? The only positive aspect of this theory is rational thinking that every deed has got its effect and it has to appear some time in future, and that everyone is free to choose what he wants to do, and this is not contrary to our common experience. It is possible that Indian thinkers of the past might have declared this theory with an additional purpose of forbidding men to do acts harmful to the society, and it seems that this purpose was served to some extent, the principle being regarded as a religious mandate. However the modern society seems to be freed from both, the merits and demerits of this theory. As already said events being the necessary and the prime quality of time, whether dilated or non-dilated, I proceed to the question whether we can have perception of time without regard to the quality. The case of external objects has been elaborately discussed in the chapter 1 and a conclusion was reached that the qualities, which depend upon the sense organs of the observing subject, are not the property of the objects at all. I proved it to you that an object separated from the bundle of qualities is a complete mystery. We do not know the thing in itself with all qualities removed from it being thrown on it by the observer himself. It has been derived and agreed by many thinkers. Just as we perceive a material object only through its qualities the time is perceived by us only through its qualities to cause events as shown above.

All that is called an event is something that happens in such a way as to become perceptible to an observer. Events include changes in properties, change in position or that in effect of matter bodies, and have role of some kind of energy in it. All this is perceptible to us through our sense organs and the mind, and as shown in chapter 1, must be classed as quality or a bunch of qualities. Perceptible qualities are observed due to the capacity of the perceiver and they really do not exist as such in the occurrence of any event. Thus events appearance as is perceptible to us is our own creation and not an attribute of the event itself. Therefore the event beyond qualities is as mysterious as the material objects, and we can not escape the conclusion that time which is evident to us through these events only must be as mysterious and as unknown as the material world. It may be that time might be a point only and we perceive it as an endless flow with events marked on it. Special theory of relativity tells us that even on material planes the time due to its property of dilation resulting from motion is incapable of being correctly measured without knowing the relative velocities, and extreme cases of minimum and maximum velocities (which does not exceed that of light) and then the time becomes so mysterious that the condition of the Universe in such cases becomes unimaginable. To top this if we take away the quality of time to be inseparably sewn to events, the mysteriousness is squared so to say. We may imagine stoppage or reversal of time and events but we can not imagine what time really is.

Chapter 3

The Space

Einstein's special theory of relativity establishes integrated behavior of space and time and as a result it is now called space time continuum and it may be felt that there is no necessity to deal with space when time has been discussed in the preceding chapter, but the space time continuum is evident when great distances and very big velocities are involved. In every day life however in absence of such conditions the space is felt as an entity with its important property to contain within it all matter and motions of matter or the waves, and it is never devoid of this quality even when it is now shown to be inseparable from time. As already discussed, although the interval between two events may be space like or time like depending upon whether the expression $c^2t^2-r^2$ has a negative value or positive value. Here c stands for velocity of light, t is the time between the two events and r is the distance between the two events. All these events appear to occur in space, and thus its quality to appear to contain events is unchanged in kind, even if the real interval may change its kind. Even when taken in ordinary sense, space has got something mysterious, although not necessarily unexplainable, in it to have attracted attention of many philosophers and physicists equally since centuries, and it is this aspect necessitating these renderings.

Like many seemingly undefined and confusing aspects inherent with the case of time, the space too can not be defined right away. Time has been dealt with in chapter 2. The child watches solid things placed here and there and it watches the change in places of things, like moving car or up going balloon and it does not have to understand that all this is happening in space. The things keep on moving and when the child grows up; his/her teachers teach him that all these moving and not moving things are in space. Space is not matter within the scientific definition of the word since it does not have weight. Space is thus imaginary thing, which we suppose to contain moving or stationary pieces of matter in it. If like many philosophers we divide all that is in the world in two parts only, that is, matter and mind, then space should fall in the category of things that are mental. It is supposed that presence of any matter does not destroy or consume the space. The presence of matter does not affect the space in anyway. If we say that this Universe is made up of matter and one thing being different from the other, then we have to look in different directions to look at the Sun and the house or a tree and if we could be as simple as the child, where is the need to assume presence of space?

But it looks that man is in the habit of creating new ideas in this old Universe and then toiling hard to define what he has assumed. He begins proving the axioms sometimes unknowingly. Well, some thinkers define space as absence of resistance to free movement of matter or bodies, but then it is a sort of negative definition of space in the sense that it defines what it hasn't got. Logically this definition is defective. The definition has to be positive as what space is. Logic tells us that definition has to describe the thing with its unique properties and not by negation of something. The definition has to be capable of making possible a systematic explanation of the subject matters it deals with.

When we see a man sitting in his room we say he is in the room. We extend this idea to the moving globes in this Universe and say that Universe must be contained in something and we call this something as space. Man is satisfied when he sees a thing being contained in some other thing. A similar assumption was made about existence of Ether when scientists were gripped by the idea that all waves should have a medium to travel through. Finally, idea of existence of Ether was discarded when it dawned upon them that light or radio waves do not require a medium to measure their velocities in relation to. God created Sun and man created idea of space to contain this burning ball, and wants further to say that if there were no space this ball could not have been where we see it.

There is more than one theory about origin of Universe. The popular Big-Bang theory says that all the matter was concentrated at a point and had infinite density in the beginning. They are not able to say why this matter was there and how it comes into existence before beginning of the Universe. If there was some matter eternally before beginning of the Universe then the Big-Bang is not really the beginning at all. Matter is as much a part of the Universe as motion and energy are and thus assuming matter before Universe is assuming Universe before its origin. And it is further said that something happened and matter exploded and threw pieces all around with enormous velocities and in course of time they took the shapes, sizes, velocities and temperatures which we now know them to have. Why an infinitely dense point piece of matter all of a sudden should be shattered to pieces? and where this enormous energy should come from is yet not clearly known. Thus those who claim the Big-Bang theory to be the right explanation of origin of Universe simply assume existence of Universe in the form of matter and energy before this event. They assume what they want to explain. If really the cause of existence of infinitely dense matter is found out and the reason for it being subjected to enormous devastating energy and the source of energy is found out, some steps could be felt to be taken to go towards where we and the Universe come from and why. Thus it should not be allowed to pass unnoticed that this theory assumes matter, energy and space to exist before the origin of the Universe, and where should the infinitely dense point matter should rest, before it is cruelly blasted by some power whose source is also unknown. Thus if space time matter and energy are fundamental ingredients of the Universe then the Big-Bang theory does not go an inch in the direction of knowing origin of Universe. However the physicists say now that the space and time both came into existence at the point of Big-Bang. They think that by such explanation the question of point of space where the dense matter was before the blast and the question since when this matter was lying there unexploded are automatically set aside. But when it is said that this peculiar singularity is the result of collapse of an earlier Universe, the question of time and space again crop up. Since how long such a cycle of big crunch or the big collapse and the big bang or the expansion has come into being is a question involving the time again even before the last Big-Bang to which the present configuration of matter in the Universe is attributed. If the Big-Bang creates the space-time continuum, the age of such continuums of the past big bangs is another question to be answered. This Big-Bang theory has as its base the general theory of relativity, which says that the space warps increasingly due to presence of bigger and bigger masses in it and is taken as truth by the present day physicists.

Now taking the space to have common meaning, forgetting for the moment the scientific theory of its origin, we find that the space is understood as a big container of things. Once we assume an entity like space to be there to contain everything which naturally is there where it is, a chain of reasoning begins to define the nature of this entity. The questions that arise are whether the space is continuous, whether it is finite or whether there are or could be any other spaces totally alien to ours? One that is; no line joining one point of our space to a point in the other space can be drawn. There are other possibilities like the space being curved, that is what we take to be a straight line extending to infinite lengths on both ends is really a curved line or may be a big circle and we never notice it. Albert Einstein in 1915 told the world that space gets warped due to presence of masses and energies in it, and what we know to be the effect of gravity is really the effect of this curving or warping. As mentioned above the physicist today believes that space came into existence when the great explosion known as the Big-Bang occurred some fifteen million years ago. The scientists do not exactly know what exactly is the volume of Universe today as yet. Let us go now in little bit of details regarding the space as is understood by common man and the philosophers.

Space and time are continuous entities and since motion is function of these two it has to be a continuous thing. I prefer to define a continuous thing as that can be divided into its smallest part without any change in its properties. The smallest part of space that is nearly a point in space retains the properties of the space. It can hold or contain a nearly point substance in it. The space is not affected by what happens within it. Space does not burn when fire is lit in it. It does not get wet when it rains. It is not electrified even if heavily charged bodies move in it. It is immune to rise in temperature even when Sun glows in it. Thus space simply contains everything without being affected by anything. And so is its smallest imaginable part. Matter however is discrete, for the smallest particles called elementary particles constituting the atom do not retain the properties of the substance of which it is a part. Electrons and protons have electrical charges on them whereas the substance may be neutral. Time is a continuous entity for its smallest possible duration still has properties of time. We can define continuity of space as follows. If a straight line could be drawn between any two points in the space, the space is continuous between those two points and if an infinite number of such pairs of points could be chosen then entire perceptible space is continuous. And this is very much true about the space we perceive for we can not imagine a pair of points which are not capable of being connected by a straight line. It is however to be remembered that presence of any piece of matter in between such points does not destroy the space occupied by it. Thus the line passes through such obstructions like matter. These two definitions indicate one and the same thing that any smallest imagined part of space is space only by its properties and that is why a straight line passing through all such smallest parts joins one point to the other.

How big is the space around us? For a frog in a well it may be equal to the volume of the well, for an illiterate tribal it could be enough to hold Sun the Moon and the canopy of starlit sky and of course his village and the places he visits. For an astronomer it is bigger than the farthest receding galaxy or some farthest star beyond identified galaxies. Thus volume of the space looks to be the product of imagination. Go on imagining bigger and bigger space and the space gets enlarged to that extent; ball of dough gets homogeneously enlarged by adding dough only to it. Thus if a body gets bigger homogeneously by adding certain thing to it, it can be validly concluded that the body must have been made up of the thing added to it. Thus it can be concluded that the space is made up of imagination only since imagination only increases its volume. Scientists however have a different idea. They say these days that although we can not find the end of the space, it is not infinite. We know that a circle drawn on a piece of paper is of finite size but to someone who travels along its periphery does not find any end to it. He goes on repeating his round without coming to the end of its circumference. Similarly any one, who travels on Earth surface for any length of time, will not come to any point where the surface terminates. This is due to the spherical shape of the Earth. We can not find the end of the two-dimensional space on a spherically curved three-dimensional body. The volume of the sphere is finite but there is no end to its surface, which is two-dimensional. It may be noted that the fact that volume of a sphere is finite is directly perceptible to an observer whose perceptions are not limited to two dimensions only. It is only when he sees the two-dimensional surface curved in third dimension forming a sphere. However if the observer has a two-dimensional perception only he can never know that the endless surface is not infinite. Similarly if there be an observer having a four dimensional perception, he can directly perceive whether three dimensional space is curved up in some fourth dimension making the endless looking space a finite entity. Physicists have now amalgamated space and time into one entity called space-time and they think that curvature in space-time makes space look endless although it is not infinite. Since man can perceive only three-dimensional things, direct verification of this speculation of scientists is not possible, but if someone should acquire capacity to look into fourth dimension a direct verification is possible. Ancient Eastern philosophers who emphasize so much on meditation claim that such powers could be acquired, but the West generally thinks this to be mysticism and does not proceed to experimental verification of the same. It is true that logic should prove a theory, but like scientific laws, laws of logic may also have limited application when enquiring about origin of Universe. When logic fails to form any theory it is intuition which suggests solutions. Hypothesis is also a kind of intuition and it takes the form of a theory when all possible corollaries drawn from the hypothesis turn out to be truly explaining the phenomenon or a thing or an abstract idea for which the theory was so earnestly sought. Avogadro's hypothesis that all gases at same temperature, pressure and of equal volumes contain equal number of molecules, has not been subjected to direct perception (for nobody would actually count the invisible molecules), but has taken the shape of a theory because all corollaries drawn from it have been found to be true in their experimental verification. The rules of logic were formulated by observing behavior of natural objects, natural phenomena, and nature of human thoughts, which creates concepts or generates knowledge about these worldly happenings. Obviously all this is after the world has been created, but when we begin applying these laws of logic to deduce the origin of the world or Universe we may land into absurdities, or we may come to a dead end and find certain mysteries impenetrable.

Could there be other spaces totally alien to the space, which we suppose to hold all perceptible matter in this Universe of ours? Boscovitch believed that there could be two or more sets of independent spaces not connected in anyway to our space and with no communication whatsoever between such kinds of spaces. The idea has further been expanded by philosopher Newton Smith who imagines two independent spaces such that no straight line of any finite length could be drawn between a point in one space to any point in the other. He has not sited any example to illustrate the point but I can site one. Space of our dreamland is a space totally alien to our waking state space and no one has ever dragged any dreamland object into the waking state world. We certainly can not draw any line between any point in our wakeful state space and any point of our dream land space. The two spaces are entirely disconnected and we have no right to say that dreamland space is false while the dream is on, just as we think our wakeful state space is not false while the wakeful state is on. While awake we say that the dream is not real but while dreaming we can equally strongly say that wakeful state world is false. After all it is same brain of ours that perceives the dream world and the wakeful state world. The only difference is that while awake our outer sense organs are sending signals to our mind of the external world, in dreams the sense organ stop functioning while the same mind works on signals sent to it in the past, recent or remote. Since not the external sense-organs, but the mind is the final perceiver, it is hard to say that one of the two perceptions is more real than the other. Some thinkers reject all that is not conceived in their waking state as unreal. Some even reject those things as unreal whose cause they can not find. Leibniz went to the extent while formulating the fourth law of logic that if there is no sufficient reason for a thing or a statement to be what it is, then its actual existence can not be real and according to him this law of sufficient reason is applicable to both, in the field of metaphysics and logic. This straight away leads us to think that if sufficient reason could not be found for existence of space or the Universe itself then it is not real at all. I shall take up this question later in this book. It is true of course that imagination of Boscovitch and of Newton Smith has to be assigned its due importance as it sprouts from this Universe and springs into some possible existence completely alien to it. That many such and even fantasy like imaginations have brought about great discoveries in science is known to all scholars of this branch of knowledge.

Is there any limit to our imagination? Obviously everyone should think that there is no such limit to human imagination. One can imagine even impossible things like a horse trotting over the convex curve of the rainbow or a particle having no mass moving on a parabolic path like a projectile on Earth, or a dead coming to life or Neptune falling on the star Alpha Centaury; Imagination is bringing back to memory the things already seen in parts or full and rearranging them in an order of our own choice to look like some real or unreal phenomenon. We take a horse rainbow and an action on walking over something from our memory bag and rearrange the things to visualize the imagined phenomenon; a boy who has never seen the sea and the Eiffel tower can not imagine the tower coming slowly out of the sea. It would be true with anyone who has not seen the ingredients of the scene to be imagined. Thus it would seem that limit of imagination is the limit of experience of perceptions in the past, but this would be an objective limit only.

Now we shall consider if there is any in-built limit of rearranging things seen in the past in human minds, which I shall call subjective limit as against the limit of the perceptions called objective limit above. A man who has experienced all worldly things and phenomena (although such a man has never existed so far) can imagine anything so to say, but a limit is invoked if we put restrictions like "being logical" and "being possible", to his wide and wild range of imagination. A horse on a rainbow is impossible. A particle without mass may be logical but impossible. Dead coming to life is impossible although superstitiously logical. A man with infinite range of worldly experiences may not be able to imagine a ball having a weight of 1 kg and 3 kg at the same time and place. He can not imagine a particle moving in a circular and a hyperbolic path at the same time and space. These are logical impossibilities. One can not imagine himself standing on his own shoulders or chasing himself on a road. Thus there is at some stage certain inherent limits set by natural laws of thoughts to the capacity of imagination. We can not imagine something coming out of nothing due to these logical limits.

However it is to be noticed that the limits get widened with inflow of new pieces of knowledge. After Einstein propounded his theory of special relativity, every scholar of physics and astronomy can very well imagine dilation of time, which was never imagined earlier. The idea must have struck Einstein as intuition when he found velocity of light to be the same whatever be the relative velocities of the observer with respect to the source of light. Intuition is not imagination. Imagination is an endeavor while intuition is an idea popping in the mind without effort.

Can we imagine a thing, which is black and non-black at the same time and place? Our logic says it is impossible. Similarly a statement of fact can not be true and untrue at the same time and place. Our worldly logic does not permit of such statements. It is simply because we can not imagine what we have never seen. Had we seen a thing to be black and non-black at the same time and place, our logic would not have incorporated law of contradiction as it is commonly called. However if somehow we reach an inference that certain thing is black and non-black in the same space-time we should be ready to partly or fully amend the law of contradiction in our system of logic.

Returning to the discussion about more than one spaces, Newton-Smith describes an imaginary land where a man from our Universe would go after eating some roots and on return he describes all things of the new world to the inhabitants of this world who know for certain by traversing the entire world that there is no other world anywhere. I do not know why a simple example of dream space could not be given where time and space both are non-unified. However the point that Newton-Smith has made is quite important that time and space could as well be non-unified. I agree with the idea that there could be different worlds having different scales and dimensions and even different spaces and times. There could be several ones absolutely disconnected. We can imagine infinite number of carvings inside an un-carved mass of stone. If a stone is broken in two parts such that one face is cut convex the other naturally becomes concave. Suppose the convex surface itself is the observer and the rest of the part is the object of perception then observer would find the object to be concave. If the stone were cut in the shape of a notch on one face and a projection on the other the observer in the form of projecting part would find the world before him as a notch and this perception is no way connected to the convex face perceived earlier. Thus we can imagine any style in which the unbroken rock mass could be broken giving rise to altogether different worlds for such observer or observers. The Universe, which we perceive, could be one of such many imagined cuts in the rock mass of the existence. The subject shall be dealt with in more details in coming chapters.

Space basically is an imagined entity to contain all matter in it and I hold that it is thus an abstract thing. It has originated from the concept that there has to be something to hold all perceptible objects. A concept is created out of many perceptions of connected things, and it is the mind, which converts perceptions into concepts. Thus concept is creation of the mind and as such it is unaffected by time or place. We can recall to our memory any concept at any time and at any place, and it comes, as it came earlier. So long as perceptions about the connected objects or phenomena do not change, the concept remains unchanged. To the question as to where or at what point the space has its origin the simple answer could be that it is the mind where the space has its birth. The warping of the space due to presence of matter or energy sources in it, as told by Einstein is equally a mental creation to suit the observed attraction between the bodies. Had Newton ever observed the deviation of light coming from planet mercury, he too could have said what Einstein has said, and this further goes to prove what I have said earlier. The concepts do not change till perceptions of the connected things change. All these theories are being coined to explain mathematically or psychologically the perceptions and the changes therein as time advances. Space, which itself is an imagination has been now loaded with another imagination of warping. Now this imaginary phenomenon is stretched to derive a conclusion that there must be a singularity in the beginning of the Universe and that Universe should end in such singularity too. In fact warping of space has simply been imagined to explain observed gravitational pull between stellar bodies, and its applicability to explain origin of Universe has never been proved. In reality even warping has not been perceived directly. No wonder if after a few years or centuries some other explanation pops up for the same age old phenomenon, making both Newton and Einstein equally a part of history. As dealt with in Chapter of 'Cause and Effect' it has been pointed out that the human perception is limited only to an observed pair of events and through repeated occurrences of such pairs the human mind infers that one event should be the cause of other. It must be kept in mind that we never perceive the actual 'Causing', or the power, which the preceding event should supposedly have and should exercise to give birth to the following event, which we commonly call the 'effect.' Science is no exception to this universal fact, and scientist's attempts to find out the reason to any phenomenon are limited by the failure of the mind in perceiving the cause directly.

In Indian philosophy space has been considered only as an imagination of the perceiving man and this when coupled with the scientific theories of today leads to very curios deductions. Physicists of today say that space has infinite number of elementary particles in it, and not only this but that the space consists of such particles although they are not directly perceptible. They can be made to manifest themselves during interactions between one elementary particle and its antiparticle as described in chapter 1. These particles make their appearance on the big universal stage of cosmic dance so to say. Thus what we consider as an empty space is not empty at all. Now if such an all-pervading entity is only an imagination, it follows that entire matter, which consists of these space-pervading particles, is also an imagination. Indian philosophers without probably deriving existence of such particles in non manifested form, declared by their intuitive knowledge as it seems, that what we call rigid matter is also an imagination. The entire existence becomes subjective thus, with the role of external objects as only to prove that the perception of them is an imagination only. The subjective-ness of the perception of Universe has been vetted by the statement of the famous physicist John Wheeler. In an exclusive interview with him made by Mirjana R. Gearhart of Cosmic Research, John wheeler said "...Stronger than the anthropic principle is what I might call the participatory principle. According to it we could not even imagine a Universe that did not somewhere and for some stretch of time contain observers because the very building materials of the Universe are these acts of observer-participancy. You wouldn't have the stuff out of which to build the Universe ......"

Chapter 4

How At All Do We Know And Who Knows Really?

The first answer to the question would be that we know because we see, we hear, we can touch things and then we read books and newspapers. True that a big part of our knowledge is through perceptions by our outer sense organs like eyes, ear, tongue, nose and the touch sensing skin; but it should be remembered that these organs do not directly generate knowledge. If we had no memory there would be no knowledge at all. When we see a car, our mind at once brings back the similar substance seen in the past from our memory bag and after finding that the substance now seen is of the same class and that the earlier seen substance was known as car, the mind concludes that the present seen thing is also a car. If we had no memory and no capacity to compare the thing in the memory with the thing in front of us, seeing even thousands of cars will generate no knowledge. Each car shall be a new thing then. Since a movie camera has no memory, no capacity to compare and no power to draw a conclusion, it gains no knowledge of the story of the film shot through it. Same applies to perceptions of other sense organs. The mind can compare sounds heard in the past with the sound being heard right now, and can recall feel of a touch in the past, and can recognize smell in the same way. Thus direct perception is the first source of knowledge.

The second source of knowledge is logical deductions. It particularly applies to those things, which we do not directly perceive but we know them through perception of some other thing or phenomenon. When we see smoke rising at a distant place we infer that there must be fire too. Something must be burning to emit the smoke and we conclude the existence of fire without perceiving it. Although it appears to be very elementary piece of logic, still the mental process involved is not that simple. We have been perceiving all the time that smoke is always associated with fire and we always use this fact as datum when we see the smoke and infer fire. We have never perceived smoke without fire. Both these facts are already stored in our memory and we apply this recollection to the phenomenon now perceived and come to the conclusion. It is pointed out here that no one has seen all cases of fire and smoke in this world. Without seeing each and every case we conclude that where there is smoke there is fire. This conclusion is by the process known as induction where a conclusion is drawn of a general nature by seeing only limited cases of the phenomenon. Conclusions by induction become wrong if even a single case is noticed say where smoke was seen to be without fire. There is one more source of indirect knowledge known as evidence. This is the knowledge supplied by persons who know a particular thing which we have not experienced ourselves. Parents, teachers and books written by scholars of various subjects are our reliable sources of knowledge. Reliability of such a source is of paramount importance. There could be defects of many kinds in information received through sources other than our direct perceptions. There can be a defect in perception. A color-blind may not be able to tell which of the colors whether green or the red he saw. One who is hard at hearing may not be able to tell the intensity of the noise he heard. A novice may mistake water for a mirage or a snake for a rope where visibility is much less. Then there could be defect in recall of his memory and he may compare the seen thing with a similar looking thing. He could take a stub of a tree to be a man in insufficiently illuminated place. Then there could be mistake in application of logic to reach a conclusion. As the humorous story of an illiterate village woman confined to her house and the nearby for years together goes, she thought that Sun rose due to her cock's crowing, for she always saw sunrise to follow the crowing of her cock, and she took it to be the reason for the Sun to rise. Of course when she hid her cock in the dark and kept it from crowing and still the Sun rose, she realized the fallacy in her logic. A single exception evaporated away the wrong reasoning. Thus information supplied to us by others can not necessarily form a part of knowledge unless the reliability of the person is established and to ascertain this the witnesses in the courts of law are cross-examined and only such part of their testimony which stands the cross –examination is accepted by the court to form the evidence. Thus all information received through other sources has to be tested and reasoned out before it is accepted as knowledge.

Is truth of the information knowledge? I know that I have yet to define knowledge. But let us examine the possible ingredients, which constitute knowledge. Suppose someone holds that the stars are nothing but the great men who lived in the past and have acquired that status in the heaven due to their good deeds. Now all of us know (thanks to the convincing statements of the scientists) that this can not be. The formation of star is now known to be the result of the heat generated due to enormous attraction between stellar dust particles, and it has nothing to do with the great souls of the past. We can not hold that the idea of the man concerned is correct and we would freely say that he has no knowledge of how stars are formed. Thus truth seems to be the important ingredient of the definition of knowledge. So now let us define knowledge as a group of ideas which are in conformity with the objective existence of things. Objective existence can be defined to be the things outside our mind and whose existence is known to us. I may add here that knowledge of the fact that hydrogen and oxygen, when chemically combined, produce water is as much knowledge of the external things as that bones and muscles interspersed with the veins and nerves make our body. Our body is an external thing for it does not incorporate the power of reasoning, which is the capacity of our mind, which in turn is a function of our brain. Even the functioning of brain has been known to quite some extent and thus the brain which is a form of matter is also a thing outside our mind which is the thinking power incarnate. Thus knowledge has to be in conformity with the objective existence and not about the real knower, for if we get knowledge about the knower, he ceases to be the knower and become one of the knowns. And if this happens then question is who the knower is then? I shall discuss this distinction between the knower and the known later on.

When we say that the mind has power to think it is necessary to know something about the means or instruments of thinking. Thinking requires help of percepts, images, concepts, symbols and certain derived formulae. Percepts are the basic ingredient that constitutes thinking. Percepts are perceptions stored in our memory, which are automatically recalled, when we perceive a similar or somewhat connected thing. Images are faint recollections of what we have perceived in the past. They are more recalled when thinking about material things. However when thinking about music or a philosophical subject images are less used and instead words are more used and in music previous percepts are frequently used.

Knowledge is a mental thing although may be about material objects. That a ball dropped from the hand falls towards the Earth, is a piece of knowledge and the thought is permanently settled in our minds without any reference to the type, size or color of the ball and without any reference to the man whose hand drops it or the height from which it is dropped. This knowledge is certainly based on our past percepts and images thereof resulting into a concept. But declaring this fact about the falling of the ball can always be without a physical object so described being present. In logic such sentences are known as propositions. A proposition has a subject and a predicate and the two are connected by a verb called copula. Any sentence is not a proposition however. A proposition is a statement which is either true or false or about which truth and falsehood can be subject of inquiry. Sentences like "Go, and bring that book to me" or "Wish you many happy returns of the day" are not propositions, for there is no expectation of them being true or false.

It is not the purpose of this book to go through rules of logic in details, but I wish that readers not already trained in logic should have some elementary idea as to what makes a proposition true and what lacuna make it false. It is called laws of thought and any deviation from these in our reasoning leads to some kind of fallacy. We need not go into the details of types of fallacies and shall only think of laws of thought here. There are four laws of thought, three of which owe their formulation to Aristotle and the fourth to the mathematician philosopher Leibniz. They are as under,

1- The law of identity, which states that if anything is A it is A. Or say if any proposition is true, it is true. It implies that the thing is not anything else but A only. A kilogram is a kilogram and nothing else.

2- The law of contradiction says that a thing A can not be B and non-B at the same time and in the same space. A ball can not be red and non-red in the same space-time.

3- The law of excluded middle says that anything must either be true or false. A proposition must either be true or false. It can not be both in the same space-time.

4- The law of sufficient reason states that if there is no sufficient reason for a thing or a statement to be what it is and not to be different from it, then its actual existence can not be real.

All these look nothing beyond common sense and in our rational thinking we utilize these laws knowingly or unknowingly. When common sense, properly sifted through sieves of reasoning by thinking over and over again on possible fallacies that may creep, unknowingly crystallizes into what is called logic. It may however be remembered that logic by itself does not necessarily take us to truth. It depends on the data upon which the logic is correctly applied. If data is erroneous the logical conclusion is bound to be wrong. We know that all men are mortal, so if John is a man he too is mortal is our conclusion. This type of thinking is called Syllogism. Syllogism tells us that if major premise; that is the statement that all men are mortal; is true, then conclusion shall be likewise.

And now let us come to the main question as to how knowledge is generated. According to Socrates, the great Greek philosopher knowledge is gained through concepts. When we see a thing or observe a thing what we get is the percept of the particular thing, but a concept is an abstraction mentally derived from common features of the class of the thing perceived. For example if we see a car say made by Toyota we get a percept of that car, but when we see a number of cars made by different manufacturers we get a concept of what a car is and this is based on noticing common features in all types of cars and also it must be able to distinguish a car from what is not a car.

Socrates believed that for obtaining perfect knowledge we must be able to define a thing very accurately; definition is, as readers will appreciate, a concept expressed in language. The definition should be able to include all common features and must distinguish the thing defined from other things. It must have common characters of the thing and also should have a distinguishing mark.

Perception may not be the right knowledge of the thing. Many times perception through our five external and sixth internal organ i.e. mind gives us illusory knowledge, like seeing a mirage, mistaking a rope for a snake, the dreams, railway track looking to converge to a point at a distance, and the apparent change in color of an object upon changing the color of its background and one may add many other cases of this nature to the list. Plato called it conjectural knowledge and rated it as the lowest type. He called all knowledge through our senses as practical knowledge. However all such knowledge is not necessarily reliable or correct, for if that was the case then there remains nothing for the teachers to teach his/her pupils. There are always two sides of a question and either of them may be taken to be correct if the other is not known, although only one of them is the truth. Further there are many phenomena which are not susceptible to direct perception.

Although human beings can be perceived, humanity can not be perceived directly. Similar is the case with many universals like liberty, equality, brotherhood and the like and these are not subjects of direct perception although these abstract things are taken as truly existing. Then there is another class of knowledge called hypothetical knowledge, which is not sensory. It includes mathematical theories. Hypothetical knowledge is arrived at by drawing logical conclusions on the basis of inductive generalizations. According to Platonic epistemology this type of knowledge acts as mediator between sensuous knowledge and the rational insight and this rational insight is rated as the highest kind of knowledge according to Platonic theory of knowledge. Rational insight gives us knowledge to form concepts or ideas. It is not the knowledge of a particular thing. It is knowledge of universals. Thus over and above the world of sense perceptions there is a transcending world of ideas and forms. The objects of sense perceptions are in space and time, but space and time do not bind the concepts that may be derived from the general features of such class of object by logical deduction. Thus concepts form and are independent sort of world which is beyond space and time. Particular space or time does not bind our concept of a car in our mind although we have seen all such cars at some place and at some time. However Plato went to the extreme when he said that there are no external objects at all and what are real are only ideas, and the objects of the external world are known only through ideas by themselves being the copy of our ideas. He just denied there being a real external world, although it is contradictory to the common experience. It is said that such an extreme thought might have become stronger in him due to him being a poet too, somewhat away from every day's mundane routine and being absorbed much in his world of thoughts. He looks sometimes oblivious of the fact that the treasure of ideas is itself derived from perceptions of the objects in this world although at later stages ideas look more and more independent of the worldly objects. His disciple the famous Aristotle differed from him in such points and disagreed on the point of existence of ideas only denying the existence of real objects of the world.

It may interest readers to know that almost all philosophers of West and East have pondered over the question of acquiring knowledge of external objects and after very deep thinking they have reached different conclusions. Each one of them took into consideration the conclusions of other philosophers of the past, recent and remote, and after applying their acute minds independently afresh on the basic question, has drawn inferences of their own. The common sense belief that all objects exist even if we do not know each of them, their qualities are their properties and do not depend on and are not affected by the knower, all object seem what they are and are what they seem and that they are known directly i.e. without anything intervening between them and the knower, is what is commonly known as Naïve Realism. This view can be endorsed by any layman not trained in philosophical or metaphysical thinking. To him the world is as it looks. His daily life goes smoothly on these assumptions and thus for him there is hardly any need to go deep into the question whether all these postulates are correct or not.

It was pointed out by John Locke (1632-1704) that although objects exist independent of their being known, the ideas about such objects depend on mind for their existence. He divided qualities of objects in two categories as pointed out in chapter 2 and he believed that mind depends completely for primary ideas upon primary qualities of the objects and these qualities are real and independent of mind. Secondary ideas which depend upon secondary qualities like color or taste are entirely dependent on perceiving mind. Secondary ideas depend on sensory channels, which receive them. According to Locke the complex ideas are created by creative power of the mind and they are formed out of many simple ideas. But since complex ideas depend on creative power of individual mind they are liable to be different or erroneous. Since the object has primary qualities like shape size or number and the mind creates secondary qualities in them, the overall perception has to depend on these two factors only and Locke believed that the real inner nature of the thing is a mystery for ever since what we perceive is limited by primary and secondary qualities only.

The striking detour from the above thinking was in the philosophy of George Berkeley (1685-1753) who said that objects, whether real things or images depend for their existence on being known. If a man does not know about an object the object will continue to exist if someone else knows it. However if no one knows the object it can continue to exist if God knows it. Thus he strongly believed that to be is to be perceived. As discussed in chapter 2 the primary and secondary qualities are not distinguishable, and both these qualities are created by the knower. All qualities depend on being known and we do not know whether there are any unknown qualities. Objects are as they seem and seem as they are only as far as our sensory experience of them is concerned. Immanuel Kant (1724-1804) has gone still more deep into the theory of knowledge (epistemology) and has analyzed functioning of mind in much detail. He divided knowledge in two broad categories as a priori knowledge and a posteriori knowledge. And said that a posteriori knowledge is gained after the experience of things, whereas a priori knowledge is something the mind already has before such experience. By a priori he meant the capacity of mind to form concepts to have idea of space and time and capacity to supply these elements to the experience and to categorize the experience or the phenomenon observed. These capacities taken together with the sensory experience generates knowledge of the thing. When we see a chair we perceive qualities as created by our minds but time and space are some things, which can not be perceived through our senses. Kant believed that these elements are supplied by our mind to the sensory perceptions and we generate the knowledge that the chair has been seen at such and such place and at such and such time. It is quite simple to say that knowledge is gained but really speaking each piece of knowledge is a judgment arrived at by us about what the thing is. Kant further says that judgment is of two types one analytical and other synthetic. The former is statement of qualities, which are already known to be possessed by the thing; like 'the sky is blue' which hardly adds to our knowledge anything new. The latter is a statement of additional information about a thing like 'the chair is broken' which adds to our knowledge, for normally what we understand by the word chair does not include its brokenness. This is additional knowledge for us about the particular chair in question. What science seeks according to Kant is synthetic a priori knowledge. Science aims at gaining synthetic knowledge or simply said new knowledge about phenomenon; and the generalization of a particular discovered phenomenon is something which is a priori or the function of the mind, which gives it the universality. In science the experiment carried out at a particular place showed that hydrogen when unites chemically with oxygen forms water, and it is the human mind which believes in the universality of the fact. This is priori knowledge according to Kant. Kant agrees that a priori knowledge differs from person to person and as such conclusions or more exactly judgments about the same phenomena may differ from man to man. Kant however does not appear to have gone into question why a priori should differ from man to man. I feel that difference in a priori knowledge from man to man may be qualitative or quantitative or both and it should be interesting to search for a cause of this difference. If these differences were assigned to nature then question simply shifts to the searching the cause why nature should be like that. Philosophers of the East have gone deep into this problem and their findings shall be discussed in chapters to come.

Thus Kant's philosophy as far as it concerns gaining knowledge is summarized as this. All phenomenal objects depend for their existence upon being known. There is no existence apart from experience. Qualities depend on their being experienced and qualities like time and space are supplied by mind to the sensory experience. If there is no error of judgment the objects seem as they are. Objects are known directly. There may be distortions in case of hasty or wrong judgment.

In the opening paragraph of this chapter I described in very simple way how we come to know what is it that we perceive. The function of mind of recalling memory and comparing the visual perception with already stocked perception may now be termed as comparing the thing seen with the idea of that thing as formed due to earlier perceptions. However the entire process of gaining knowledge is not that simple and philosophers have different opinions about it. The first step in getting knowledge of external world (that is the one outside our minds, to define it simply) is to have a sensation through our sensory organs. This sensation by itself does not generate any knowledge. As told above, the time and space elements are supplied by the mind to the sensation, and the element of belief that the thing exists has also to be supplied by the mind, for there is nothing in our sensation itself which conveys the idea that the thing exists. Thus time, space and existence are supplied by the mind and then a percept is formed. Many such percepts give rise to concept about a class of things.

Now we come to the question whether the things which we can not or do not perceive through our six sense organs exist or not? The first reply which common sense demand should be that things exist and we may know them or we may not know them. Take for example the existence of Neptune or Pluto orbiting round the Sun. About hundred years ago we did not know that such planets exist. Now we do know about them and their so many properties like their diameters their average temperatures their periods of revolutions round the Sun and many other such things. Now it would be most unscientific to say that these planets did not exist at all some hundred years ago and that they came into existence the moment they were detected by the noted astronomers. Our knowledge of external things is limited by the capacities of our six organs beyond which we can not go. Suppose we had no organ of sight that is our eyes. Imagine how would have been our perceptions of the world. There would have been no colors no light no visual idea about distances and so on. Imagine if we did not have audio organ as well; then our world would have been without any noise at all. And if in addition we did not have the sense of smell we would not have recognized presence of flowers and even spices. How can we say that the five external sense organs and the internal organ which we call mind is perfect enough and all the necessary organs which one should have to gain full knowledge of the external as well as the world of thoughts? It is thus wrong to think that the things which we do not know of do not exist. However it is equally wrong to think that things we do not know about always exist. They may or they may not. But if we could find a logical chain between what we know and what we can not or do not know, it will establish the existence of what we do not or can not know. If it is said that a thing in itself is beyond all perceptible properties, there has to be logical derivation to that effect, in absence of which one has to agree with the Neo realists view that a substance or a thing is nothing but a bundle of qualities, and the name given to the thing does not convey anything beyond the qualities. The name they say is a precise form of describing all qualities which may be too many to be conveyed individually. It shall be interesting to know the arguments put on by those who believe in a separate existence of the thing in itself beyond all qualities.

I shall again return to the problem of knower and the known which was only peripherally touched earlier in this chapter. When we know about a table or a car, we are the knower and the table or the car is the known, and this looks very simple and clear. But the problem gets more and more complicated as we think deeper than this. All external things of which we have the knowledge are the known ones and we are the knower. When I know about my legs or hand, the leg or the hand become the known and I am the knower, but here the hand or the leg is treated as external object although they are the parts of my body. In such a case we have to say that it is my mind which is the knower and all parts of my body are the known. Brain which should be the seat of mind is also not an unknown thing. Neuro surgeons know about brain much more than we know about our legs. Thus brain has to be treated just as an external object of knowledge. Now we come to mind which is the function of the brain just as the locomotion is the functional result of a locomotive. Psychologists have analyzed the working of our mind in great details, and there are more than enough books detailing how mind works in varied states and conditions, from a normal man to mad and insane or lunatics. Thus to a psychologists, mind which is a function of brain, is almost fully known. Thus it seems that even mind too has to slip away from the category of the knower and be among the things known. Then who should be the knower is still not known. And really why should the knower be the known? For then the distinction between the knower and the known should vanish which is absurd. If the distinction has to exist then we shall have to say that the knower can never become known for then he ceases to be the knower. It is not correct to think that some part of the mind is the knower of the function of remaining part of it, for then the question as to who knows that a part of mind knows the other part can not be replied at all. Thus if a distinction between the knower and the known is to be honored, we can not know the knower at all, shall have to be logically accepted. If we are not prepared to recognize the distinction between knower and the known, we shall have to say that knower and the known are one and the same thing, and then shall end all practical duality in the world and we shall have no right to say that knowledge is a real entity so far as external world (as is normally called being outside our mind) is supposed to exist in relation to us. Thus it shall have to be concluded that the real knower is beyond our minds but is not beyond ourselves for it is we who say that we know about this or that thing including our brain and mind too in such known things.

Chapter 5

How This Creation After All

The dream is an example of something being created with no material to work upon. There is no dream space, which precedes the rest of the dream. No one has ever dreamt himself creating the dream scenes out of such a dream space. There is no order in this creation. Dream space does not precede the dream matter or the dream time or the dream energy. All of these seem to crop up simultaneously. If we presume the same principle to apply to the creation of the Universe, we have to imagine someone say God to dream this entire Universe as it looks to us. We are the dream people of God, and all objects, phenomena and all energies are His dream creation. God himself is the material out of which the entire panorama of his great dream looks to have been created. His is the only mind that imagines logical thinking or reasoning as a priori to all beings; particularly the human ones. What we perceive as worldly objects is all His creation. Our perception is also His imagination since we seem to perceive only as a part of His dream action. Readers may feel here that this is something what Berkeley used to believe, but there is one difference. Berkeley considered entire creation as real one and not a dream world of God. Well, when we are dreaming the dream world looks as real as the wakeful state world. Thus even if we imagine the Universe as God's dream it does not make any difference so long as this great dream is on. It looks as real as a real Universe could be. Now I come to some subtler points regarding this great dream of God if that is so. When we dream talking to someone say 'A', thinking while we talk is known to the dreaming ourselves. We do not know the process of thinking in the mind of 'A' when he is talking to us in dreams. We can not enter in the mind of 'A' and begin thinking and talking to us. I read a very interesting story about dreams. A princess was telling her maid about the horrible dream she had the night before. She dreamt that a prince from some other country grabbed her by hand and pulled her on his horse's back and began galloping away to a place unknown to her. The stunned princess asked as to where he was taking her after all, and the prince looking in her eyes and snapped "You should know better for it is your dream, not mine".

Thus first person pronoun 'I' can only be used by the dreaming us, and not by 'A' that is the person to whom 'I' in dream is talking to. We can not become 'A' and talk to ourselves. In this Universe each one has the right to become the first person and do the thinking and talking. It is the sleeping I who is dreaming and in the dream also he is the 'I' of dreams. In the waking state as well we can not enter others mind and be aware of their process of thinking. We are conscious only of our own thinking, perceptions, knowledge and emotions. Thus if this dream theory is correct then it is not Gods dream. It is our own dream. The reply to the question who dreams? Is that the person who feels himself to be the 'I' of the dream. In this respect it would be very interesting to note that Eastern philosophers to whom the scriptures 'Vedas' and the 'Upanishads' are attributed have declared that there is absolutely no difference between the Absolute and our own self but for the ignorance of this truth. If we think in those lines the reason why everybody plays 'I' in this dream-like-existence becomes at once very clear. Upanishads declare that there is only one 'I' and that it is this 'I' manifests itself as multiplicity of minds, plurality of people, variety of objects seemingly external to us. This explains how everyone thinks himself to be the first person and the witness of all phenomena that occur in this Universe. However the reason why this fact should not be self-evident is not very plain and simple. Why a man does not know from his birth or in his young age the fact that he is identically the Absolute Itself is not very clearly explained in Upanishads, and the authors have taken resort to Illusion or conscience to answer this. Why this illusion pesters everybody from his birth is also not convincingly answered in Vedas and Upanishads and commentaries on these by learned and the seers of the East. Well, so long as the entire truth is not very well known as a writing on the wall, any one philosophizing in this regard has to take resort to at least one such unexplainable entity which is in result responsible for this existence, and probably best philosophy would be the one which supposes or expects least number of functions in such entity. If it be said that the cause of rains is some goddess who creates clouds, brings them above our heads, then at her Will asks rains to begin, then except that it finally rains; all other things or phenomena are hidden from us. Now it has to be considered superior theory to this when it is said that Sun causes seas and oceans to emit vapors which reaching high up become clouds and then the wind drives them in its direction to reach areas where we are, and then after percentage of humidity reaches a limit the lowering of temperature results into condensation of clouds resulting in rains. The latter theory assumes only one unknown factor that is the reason why water should change into vapors due to Sun. That philosophy is certainly superior which has least unknown entities. The primitive man must have taken all physical phenomena as wonders of the world while today sciences dealing in various branches of knowledge have reduced the most complicated phenomena of the past to very simple combinations of very simple elementary facts. Of course as already said some simple facts like why electrons or protons should have electric charge or why mass and energy should be inter-convertible is still attributed to something unknown or to Gods play. Further development of sciences will further reduce the Gods role, but to date, it can not be said for certain that a time will come when man would know everything about the existence as a whole. Even if we completely come to know 'How' of the Universe and are able to predict what should happen next moment science can not know 'Why' of the Universe if it limits itself by what is called scientific methods of today. Science shall have to re-orient and redefine itself to reach so far unfathomable depths.

Chapter 6

Cause And Effect

With this chain of cause and effect we seem to be all tied up to the existence. We can never escape this. Everything that happens in this Universe has a cause of which the happening is the effect. Changes do not occur at random. We may know the cause or we may not, all changes follow a system in time which we call chain of cause and effect. The cause of day and night is the duration of rotation of the Earth and the cause of winter and summer is the orbiting of Earth with a definite constant inclination around the Sun. The cause of today's situation is the yesterday's situation speaking generally, without reference to any special happening. Chair or a table is the effect of material that is wood, a design as to its dimensions of various parts and the skilled labor put into it. Further the effect itself becomes the cause of further changes. Thus every cause has an effect and every effect in turn is the cause of the next change. This chain of causation seems to be beginning-less, if each effect is to be the cause of the next change and howsoever one may try to extend the chain backwards, he is not going to find the first cause or the first effect. It is like trying to know which came first, the hen or the egg. Many philosophers who claim or presume that the world had a beginning think that God must be the first cause behind this creation and present situation is the effect of the very long chain of effects and causes since then. Physicists, who can not explain why initially an infinitely dense point matter existed and why this mass was cruelly blasted, end up bringing in God as the answer to the ultimate 'Why'. If the beginning of the Universe is without a cause, then it is possible that all subsequent events may be similarly without relation of cause and effect and it may be only our idea based on observations of occurrences that a relation of cause and effect must exist. If God is the first cause of the first event of this Universe coming into existence, then no one can stop Him from being cause of the subsequent events directly thus eliminating the necessity of the preceding event to act as a cause of the subsequent event. Everything is done by God directly then and there is such thinking prevalent among many religious thinkers indeed.

The cause of a particular quantity of water turning into steam is application of heat to the water. But this is not enough. That it is the property of water molecules to turn into steam molecules is just telling the fact of water turning into steam in the other way. The molecules of water and that of steam are identical structurally but the only difference is in their mobility. Heat increases the mobility and hence the velocity of the molecules. The cause should exactly be sufficient to explain the effect. Water at ordinary room temperature when heated becomes hot until it starts boiling. And then steam issues. If water did not have this property, no steam would have been produced howsoever we might heat the water. The first cause of water becoming steam is the property of water to change from liquid state to gaseous state at a particular temperature and corresponding pressure (because it boils at different temperatures if pressure over it is changed), and the second cause is that such a temperature was reached by application of heat. We know that simply reaching the temperature does not cause entire quantity of water to convert itself into steam. Heat has to be continued to be applied till the entire quantity of water converts into steam and disappears.

The cause of rusting of iron in damp weather is also two-fold. First is that atoms of iron have got affinity towards combining with oxygen atoms of the air in presence of water or water vapor, and the second is creation of the situation which is conducive for this reaction to begin. It will be noticed that the effect is already contained in the cause itself. Affinity of iron towards combining chemically with oxygen in a particular condition is the effect already contained in the cause. This is true for all chemical reactions and all physical changes seen around us. When a statue is carved out of a big stone, it needs only an act of carving by a sculptor to change the stone into the desired statue. The sculptor chips off skillfully the undesired part of the stone mass and lets the balance part of the stone conforming to the desired figure remain as it is. Thus we may say that the statue is already contained in the stone and it only needed to be revealed by the skill of sculptor on it. And then why limit us to the particular figure only? In fact a mass of stone already contains an infinite number of figures, and it only depends on the desire of the sculptor that which one of these figures he carves out. We may make here slight difference in the meaning of words 'cause' and 'reason'. We use the word reason for some mental action or some concept or past percept brought into play consciously or unconsciously in the study of the phenomenon of cause and effect. We may say that the statue was already in the stone as effect is contained in the cause, but the reason for its appearance is the skill of the sculptor. The cause of rusting lies in the property of iron to combine chemically with oxygen and the reason why a piece of iron rusted was that it was left in the situation conducive of that reaction. It is not always that every change is brought about by known reasons. There are many phenomena for which material cause may be known but reasons are not yet known. Diseases like cancer may fall in such category where material cause is known but exact reason for the chain of action may not be clearly known by which the cause culminates into the observed effect. But we believe that some day all reasons for all the changes in this Universe shall be known, and it is on this hope that science exists and develops. The reason for Earth going round the Sun is the gravitational pull, which is a concept, logically developed by the scientists. Even the modern concept that space-time gets curved or warped due to presence of masses and energies in the space has also been derived from past percepts, and is put forth as a reason to explain what we consider to be the effect of gravitational pull. The reason is some times tailored to suit the effect, although it may be incapable of being logically derived from our percepts. However this tailoring is not intuition within the exact definition of the word.

Although the effect is already contained in the cause, the two things may not be simultaneous. Statue in the rock mass is invisible till carved out. The colors of the spectrum which are already contained in the ray of white light are not visible till a prism is interposed between the eye and the source of light. It looks as if the time separates effect from the cause, or in many other cases space may be the difference between them. A ball falling freely under the force of gravity will show its fallen state after it has traveled some distance, which in turn has taken corresponding time. When we say that today's world is the cause of tomorrows world, a day has to pass in between and the Earth would have traveled miles in its orbit around the Sun by that time. The question whether time causes changes or the changes give rise to the notion of time has been discussed in details in chapter 3. Time and space both are not substances, and their existence is only in the form of concepts that are derived from past percepts. If matter is the only reality then time and space both are unreal, but fortunately it is not so, because motion of bodies in this Universe is a function of time and distance and although we may not be able to imagine a real point in space or a real moment in this eternal entity known as time, we are definitely able to imagine a finite distance traveled in a finite time.

It would seem that for an effect to occur, we require some material or substance to either work upon or to be the source of perception of the cause or the effect or any intermediate stage and then the energy to bring about the change, and important of all is the desire or will or design to have the effect. We may classify the material or the energy as the basic cause but the will or desire and the intermediate steps in the will also constitute the reasons for that effect. It is not always that reason is or may be clearly known in case of many phenomena in nature, but usually cause may be known as far as possible. When a carpenter makes a wooden chair, he has the wood, nails and screws as the material, the skilled labor he performs on the wood is the cause and the will or the design according to which he proceeds with his works may be called the reason behind the making of the chair, although speaking generally we don't make much difference between the reason and the cause and these two words are often used to mean the same. After making the legs of the chair the carpenter proceeds to fix the board for the seat or for the back according to the design and these intermediate things is what is meant by intermediate steps to be included in the word 'reason'. All man-made things have similar procedure and steps in their making; house or a motor car or a rocket are similar things as far as these steps in their making are concerned. However the natural phenomena are not that easy to explain. Not only that, but in case of man-made things too, it is difficult to explain the fundamental properties of the material used, for these properties which permit the change of shape or state are natural and still we proceed explaining things assuming that these fundamentals shall not be inquired into. It is not easy really to search an answer to the ultimate 'Why' of things.

Indian philosophers of the past claimed that the cause and the effect are identical, and for this they choose an earthen jar as an example and say that jar is nothing but Earth only of which it is made. They separate the form of a thing from the matter of the thing and say that it is only difference in shape of the Earth, which separate the visible form of the two and declare that the two things are not really different. The effect is the cause itself. I think this view is not acceptable as it is; so far as everyday perceptible changes are concerned. The lump of earth had under gone change in shape due to the skilled labor put in by the potter and the heat that has been supplied to the shaped earth so as to make it hard enough. This input of labor and energy in the form of heat can be termed as efficient cause as against the wet earth named as the material cause. Thus the wet earth of the required quality plus the skilled labor done on it plus the heat treatment, turn the earth into a pitcher or a jar. When we add to this the will to make the pitcher or the desire to have it made, the entire reason is known. Thus for all man-made things the process can be split in four distinct features.**1**, the plan; **2**, the material; **3**, the application of physical energy and **4,** the skilled or unskilled labor. Out of the above the second one can be termed as the material cause and the rest three can be summed up as efficient cause. It is not always that all the four ingredients are brought into play as in case of rusting or dilapidation of an old building, skilled or unskilled labor may not be required. It also does not need a human plan to effect the change and may not need artificial application of any energy either. Now we consider certain natural phenomena and find out whether cause or the reason is perfectly known or is capable of being known. The cause of one's catching common cold or say influenza is known to be certain half living things called virus, and this when enters the mans respiratory system and becomes stable and active, the decease results. The conditions in which such an attack of the virus is possible are also known. It is also known that the virus may pass through exhaling of an affected person and may attack the nearby person if he has a low resistance at the moment. Why should virus cause symptoms like fever is also known, but why should such virus have this property of causing the disease is not known. Why should such a virus be present on this globe is also not clearly known. When and how the first virus was created is also not known. It would then appear that the entire cause and reason for catching such a disease is not known yet, but as medical science advances many of such 'Whys' shall be answered.

Aristotle distinguishes between four types of causes, which are **1,** the material cause **2,** the efficient cause **3,** the formal cause and **4,** the final cause. According to him every object has two aspects; the material and the form. All objects that exist must have these two aspects invariably. Returning back to the example of making of a chair, the wood and the nails etc is the material cause, the carpenter is the efficient cause, the pre-decided shape or the form according to which the chair is made is the formal cause and the purpose for which it was made is the final cause. However in the natural phenomena as rusting of the iron rod the final cause may not be ascertainable although we may say that nature wants it to happen. The formal cause can also be loosely attributed to the nature. The efficient cause may be the property of iron atoms to combine with oxygen in the conditions favorable for the reaction to take place.

The above discussion leads us to think that we might now be more exact in defining the cause and the effect thereof. Since the chain of cause and effect seems to extend infinitely in both the directions of the present tense we may say now that cause and effect are relative to each other, meaning thereby that the effect 'x' has got a cause "w" and the effect 'x' becomes the cause for an effect 'y' which in turn is the cause of effect 'z' and similarly proceeding backwards we find that the cause 'w' is the effect of another preceding cause 'v' and which in turn is the effect of a cause 'u' and so on. Thus a cause is not a cause in all contexts and effect is not an effect in all contexts. It is also seen that cause either precedes or is simultaneous with the effect. It can never succeed the effect and it is here that the time factor comes into play. Simultaneous with the birth of his child a man becomes father. The cause of his being called a father is the birth of the child. This of course is not the result of any material change in him. Water when heated sufficiently evaporates. However we can sight no worldly example where effect precedes its cause. Scientifically the cause may be defined as the entire aggregate of conditions of the circumstance requisite to the effect. Or as Mill says it is the sum total of the conditions, positive or negative taken together requisite to the effect.

Philosophers of the Sankhya school of the Eastern philosophy distinguish between the material and the efficient cause by saying that the material cause enters the effect and the efficient cause acts from outside. Although the effect is already hidden in the material cause the efficient cause is required to manifest it. According to Sankhya the Universe is made up of various things which are the effects, and they have their origin in the material cause in this chain of cause and effect. So there has to be a fundamental cause for this Universe. They name the fundamental cause of all limited things in the world as 'Prakrati'. They say that worldly objects which are born and die, which are created and destroyed, which are thus limited within space-time continuum have the Prakrati which is beginning-less and end-less as their fundamental cause. The Universe is an effect and it must have a cause in which it lies unmanifested and this unmanifested cause they call Prakrati.

Now I shall turn to still more fundamental problem. The question is, on what basis we believe that everything that happens in this Universe, has a cause? The mathematician philosopher Leibniz believed this so strongly that he concluded that if no sufficient reason is found for a thing or an event, the thing or the event can not be real. When we see an iron rod lying rusted in a shallow damp ditch, we begin searching for the cause of that particular state of rust and we find out that iron in the presence of moisture and air must have got on its surface, oxides of iron, and this appearance is what we call rusting. This is what science has been telling us for certain. But the question is why we started with the belief that every state of things has got a cause? How this idea got impregnated in our minds that every situation has got a cause behind it? We may hastily say that it is our past experience but it shall become clear that this instantaneous reply is most crude when we start defining what experience itself is. To perceive a pair of cause and effect we have to have sensation and perception of at least two things, one of them to stand as cause and the other for the effect. All the qualities of the first thing are generated by the knower himself as pointed out in chapter 1, and the knower is never able to know **What** is not a subject of perception at all. The perception of the other thing likewise is restricted to this necessary belief that we can not know what is not perceived by us. Now for the second thing to become effect of the first, we should be able to perceive some power, which makes the first thing as the cause of the other. But since we are not able to perceive anything but qualities of the object, created as a result of effect, we never perceive anything like a hidden power or the capacity of the first thing to give birth to the second thing. Thus the phenomenal external objects do not supply idea of cause and effect to us. The causality can not be deduced by observing the effect or the next perception that we believe to be the effect of the first one. We can never know the cause by simply observing what we call the effect. But at the same time we believe in the chain of cause and effect so much so that a common man will simply call a person insane who denies existence of causality in a general sense. Let us examine then the reason of this common belief.

Perception of pair of events can not give the idea of cause and effect. Even minutest observation or sharpest mental exercise is futile. Thus it has to be concluded that the idea of causality is supplied by our minds only. The reason that all of us feel that there is always a cause to every effect must be the inborn, a priori knowledge in the minds, which tend us to search for reason of all phenomena in this Universe. And when we find that science has found out causes to a number of happening in this Universe and they are uniformly operative, independent of where and when they happen, we have all rights to conclude that there is an important connection between a priori knowledge in our minds and the behavior of the external phenomenal objects of this Universe. And why should this not be? Are we not as much a part of the Universe as are these seeming external objects? There is some underlying existence common to all things and phenomena. I shall re-enter into this aspect later on but a word of caution will be proper to be given here. There are many new discoveries in science, which have proved the earlier theories to be wrong although at the time these earlier discoveries were made, they perfectly fitted in the sockets of cause and effect in the minds of the earlier scientists. Theory of relativity has proved Newtonian concepts of time and space to be wrong is a glaring example of this. And that there are many such examples in not only physics or chemistry but also in almost all sciences like medicine, botany, biology, engineering and so on, is indicative of the fact that a priori knowledge may not always depict or dictate external world. There might not be any cause and effect phenomena at all in this phenomenal Universe and the idea could be only a belief that such a priori knowledge exists. Universe may be a simple flow of changing scenes without any one of them being the cause of the next and it may be just a vain pursuit of our mind to correlate these changes due to what we call a seeming a priori knowledge. This could be the reason why we do not know the first so called cause of this entire creation as pointed out by me in the opening paragraph of this chapter, and it was found convenient to attribute this capacity of having a causeless result as the beginning of the Universe to God (without taking his permission to make such statements). . Sir Isaac Newton was no ordinary man of those days and his so-called a priori knowledge about time space and causality went wrong and we have no reasons to believe that he was the last physicist of the kind.

It is certain that since the idea of one happening being the cause of the other, has not been supplied by our perceptions, it is always supplied by our minds. Our minds may deceive us. As the story goes, an old village lady used to think that Sun rose because her cock crowed in the morning. We have no reason to assume that even our trained minds supply correct causality to any observed phenomena. We could be just like the village lady, and scientists are no exceptions. Even if we find certain two happenings always associated with each other, we may not know the truth as to whether they are connected as cause and effect or not.

Philosopher David Hume says that a priori reasoning is not the sound method to arrive at the causal relations. Without experience no one can be able to deduce the causality. No amount of arguments can give us the knowledge of how things function. In fact what we call cause and effect is a relation, which can be induced only on the basis of past experiences. He says, it is only in mathematics that we find different ideas interrelated invariably. Facts, which we perceive as the external objects or phenomena, do not have such relationship as mathematics. We can not deduce one fact by observing the other even if we believe that there is cause and effect relationship between the two, whereas in mathematics we certainly can do this. It can be deduced in mathematics by studying the equilateral triangle that all three angles must be equal to one another. However if one takes such corollaries as the effect of a cause then of course I differ. That each one of the three angles equal 60 degrees is not a new or a subsequent fact. It was right there the moment we imagined and considered an equilateral triangle. All such facts like trigonometric relations come into play as soon as an equilateral triangle is imagined. The only thing that happens when we prove geometrically a fact is that we become conscious of the already present conclusion by following the line of proof. It is not a new happening as an effect of the imagining a triangle or any other figure. If (a+b) when squared begets $a^2+2ab+b^2$, this second expression is not a new fact, as it is already there by virtue of existence of (a+b). Thus I hold that there is no cause and effect chain in mathematical formulae at all. All that is there exists simultaneously while a student becomes conscious of these facts one by one as he advances in his studies. Mathematics is that way an analytical knowledge in which various logical predicates can be added to one subject. Hume of course is right in maintaining that ideas in mathematics are inter related while facts do not possess such relationship. Even when we desire to raise our hand above our heads and hand goes up, the two experiences one of the volition and other hand going up are recorded in our minds but, although we find that the two experiences followed one after the other, we can not observe anything which relates the two observations as cause and effect. According to Hume we generate the knowledge of cause and effect by habit and custom only. The mind has the habit of expecting another occurrence when one has taken place and this habit is formed on the basis of past experiences and this is what we call causation although logically the two happenings do not have any interrelation to that effect.

Now I come to an interesting example of attributing the observed effect to a wrong cause. Albert Einstein explained how the observed facts might lead us to wrong conclusion about its cause. The force of gravity, which was so very reliably attributed to the property of mass by Sir Isaac Newton, was shown by Einstein, to be due to curvature or distorting or warping of the space time continuum due to presence of what we call masses and energies in this Universe. Physicists explain the theory by a thought experiment as they call it. It runs something like this. Suppose there is an elevator as large in size as about one tenth or one twentieth of the Earth's circumference and it is falling on the Earth since a long time and that it contains physicists who were born and brought up in this windowless falling elevator. Since it is falling freely under gravity the persons in the elevator do not experience any force of gravity and there position resembles the position of astronauts staying freely outside their space shuttle. The things left by them continue to be there as they left them. They do not fall down. Suppose two balls have been placed by the physicists inside the elevator at a very large distance, say about a thousand kilometers apart, then as the elevators travels down, the balls would be seen having come closer due to the fact that each one of them is attracted towards center of the Earth and as such their paths shall converge as they come closer to the Earth's surface. Now the physicists inside the great elevator shall attribute this phenomenon to the attraction between the balls, whereas there is no force of attraction that exists between them. Thus even the great brains like Newton misjudged the phenomenon of seeming attraction between bodies. Thus we must come out of the thought that the physical world behaves according to what our brains conclude. Not at all. Even the greatest brains may lead us to wrong conclusions. All this separates the actual world from the world we think it to be.

However this does not indicate death of Logic. Logic is still alive and can show logically under what circumstances the conclusions drawn may go wrong. But at the same time it has to be noted that logic has never claimed that it would always lead you to the truth. It simply states rule how thinking proceeds in human minds. And all these rules depend upon worldly examples of thinking. A black and a non-black thing can not exist in the same space-time is a common rule of logic, but as human mind grasps more and more of the so far unrevealed world, a stage may come when logic shall have to admit of a black and non black thing at the same space time and shall have its one of the famous rule stand amended. Well no harm by that if knowledge is considered more important than the Aristotelian way of thinking of some 2500 years ago.

George Berkeley believed that God is the sole cause of the world and it is He who created individual souls and ideas. Berkeley holds that it is wrong to say that fire creates heat; Fire and heat have no such causal relation within themselves. In fact heat should be taken as sign of fire; He had no objection when a common man said that fire creates heat because he has always seen heat generated after the fire. Berkeley believed that all causality is in God. Fire is antecedent and heat is the consequent and this is so because God has willed that heat should occur where there is fire. The real cause of this causality is will of God. It is due to God's will that we find so many happenings connected with cause and effect relationship. Berkeley believed that it was easier to explain the behavior of the Universe if it is accepted that everything happens in this world because God wills it to happen. It is He who creates the idea of causality in our minds, although there may not be any such relations in the things or happenings in themselves. Berkeley was a religious man, and although his concept looks more religious than philosophical, it would seem that Hume derived his theory from Berkeley's idea that the causality is not present in the observed phenomena but is a creation of mind. Immanuel Kant held that causality is nothing but a category of thought. All knowledge we possess, comes to us after having been processed by the mental structure, and we cannot say for certain that causality exists in the objects of the external world, because man's knowledge is limited by his mental structure. It is clear from above that Berkeley's concept is more based on intuition than on a rigorous logical analysis or synthesis, but it must not be forgotten at the same time that even in science many times intuitions have led to great truths.

We may defer finding replies to the questions, whether causality is existent in the external world or not; or whether it is entirely supplied by the mind, till I reach the last two chapters of this book. I hold that these questions involve much deeper consideration, for these involve existence or otherwise of causality in the ideas unlike causality of the external world. Causality in the external world can not be a matter of experience, but causality between two ideas always associated, is a matter of our own internal experience. In external world we may think that causality exists between two happenings while really it may not be there, while in the internal world of thoughts we can directly experience the causality between two thoughts, and this experience must not be compared with the external experience, for the latter gives the knowledge of the qualities but not the knowledge of the thing in itself. The internal experience of the thoughts gives us knowledge of the thought itself and not the knowledge of the quality of thoughts.

As pointed out above the causality is never a subject of human perception. When dilute sulphuric acid is brought into contact with zinc, the result is formation of a compound known as zinc sulphate and hydrogen evolves. This is explained on the basis of dissociation of the acid into ions, which are attracted towards the oppositely charged ones resulting in the products of this reaction. Ions could be explained on the basis of atomic structure and shifting of electrons according to the chemical valency of the elements and so on. However we can not perceive the cause as to why oppositely charged ions should attract each other to combine into a compound. We perceive the result but not the real cause. We do not perceive the power, which cause this reaction although scientists finish it by saying that it is the property of the substances, elements, radicals, ions or electron to behave so. What is the power that gave these properties to the reacting ingredients is never perceived by us although scientist may further find some simpler facts in future to explain the reaction, but they would never know the ultimate why of this chemical change. Explaining complicated facts by simpler facts does not lead us anywhere when the cause of the simplest fact in this chain of explanation can not be perceived. It thus remains a fact that we do not perceive the power that causes changes in this Universe and yet we believe that every change must have a cause. We can only say that there is some power that has triggered this series of changes and whether one is the cause of other or of the subsequent event or not. If the Universe is a sequence of unconnected events so far as causality is concerned, there is no point in attributing the first event to have been caused by God and to top this there may not have been a first event at all, if the existence is beginning-less.

Thus what we call convincing reasoning is nothing but a result of first splitting up exhaustively a complicated phenomenon into elementary pairs of events, which are a part of experience of both the parties to an argument, and then adding up skillfully such pairs, till the complicated phenomenon logically looks to fully depend on such an addition, and the resultant cause and effect of such addition looks to be the cause of the phenomenon in question. After all, all sciences have to develop on what our minds think to be the logical connection between observations and theories explaining them. There is no alternative (except what is believed to be clairvoyance, which is not common to all) It must also be noticed that all convincing may not lead to the truth. Ptolemy must have convinced the then world that Sun and the planets go round the Earth. This was however not the truth.

Chapter 7

Forces And Energy

Scientists have found out that there are only four types of forces that govern motion and equilibrium of all stellar and terrestrial objects. Planets do not spiral around and fall into the Sun due to balancing of centripetal force, caused due to motion and the gravitational force, that prevents the planets from flying away due to the centripetal force. The positively charged body is attracted towards negatively charged body due to electromagnetic force.

As simply defined, energy is capacity to do work. If a man feels energetic it means that he can do more work. In science however the definition although remains the same, the energy has been quantified depending upon what amount of physical work can be done in a particular quantity of energy. The work is defined as a product of force and the distance through which a body travels against it. If a weight of one gram is lifted by one centimeter the work done is measured as one gram-centimeter or erg. Thus unit of energy is identical with that of work it is able to do. A body has at a time two types of physical energies; one is potential energy and other is kinetic energy, by virtue of which it being at rest or in motion respectively. Potential energy is by virtue of its position with respect to Earth or any other planet or star on which it is situated, and is measured as its weight on that planet or star multiplied by its distance from the center of the planet where its gravitational force is supposed to act from. Thus on Earth, one gram of weight has got a potential energy of one gram multiplied by the acceleration due to gravity, further multiplied by the radius of the Earth. However for terrestrial, simple problems as in engineering or hydraulics, it is customary to measure the potential energy with respect to the Earth's surface. Thus a mass of one gram placed at a height of 100 centimeters from Earth's surface has got a potential energy of I multiplied by 981 multiplied by 100 where the acceleration due to gravity is 981 cm /second/second. The unit of the energy will be dyne centimeter. Thus the energy in the above example will be 98100 dyne centimeter. The kinetic energy of a body of mass M grams, which is moving with a velocity of V cm per second, will be 1/2 M multiplied by square of the velocity that is V^2. The two types of energies are inter-convertible and when a body freely falls under the gravity of the Earth, the potential energy by virtue of its earlier position converts itself into kinetic energy due to the velocity it gains due to its fall.

But these are not the only types of energies. The electrical energy is measured as a product of the voltage and the current. The energy by virtue of heat is measured in calories, the energy transmitted by light is either measured in calories due to its conversion into heat or by the Planks equation e= cf. where 'c' is a constant and the 'f' stands for the frequency of the radiation. In fact all types of physical energies can be converted into their heat equivalents and can be measured as calories or by their physical equivalents and can be measured as gm and Centimeter. Atomic energy, Nuclear energy, energies due to chemical reactions are all thus measurable in terms of physical or more conveniently as heat equivalents. All types of energies are inter-convertible. Even the energy to be available from food intake is measured in calories.

Thus we find that energy in whatever form it is supposed to exist must have matter in the Universe for its manifestation. Energy needs matter, which is the stuff it works upon. Energy and matter are inter-convertible but if entire energy converts itself into matter, the matter would be infinitely dense, for intermolecular and intro-atomic movement should also cease when there is no energy at all. Thus in only-matter situation, there wont be left any energy to once again blast it into Universe as Big–Bang theory propounds. In case of no-matter stage, there shall be no manifestation of energy and condition would be as if there is neither energy nor matter. This condition should not be allowed to look similar to Sankhya's 'Prakrati', for there won't be left any 'Purusha' or the knower or the enjoyer of the 'Prakrati', to know the unmanifested energy subsequently changing into the matter and have the Universe.

The reality of matter is not ascertainable on the basis of qualities only. If all qualities including its mass are removed from any object then whatever remains can not be subject of energy to work upon, for energy must be able to change the position if mass internally, as Leibniz believed, it should be able to show its primary and secondary activities. Thus energy can not be more real than the matter. If matter is unreal, the idea of energy is equally false.

As scientists say, thinking also involves consumption of energy as far as physiological function of brain is concerned. Thinking is a process of consuming energy in one form and its manifestation in other forms like memory, judgment, desires and impulses; urging physical activities in the interest of self-protection and having pleasure. If brain is deprived of its supply of oxygen, glucose and the like, the functions of brain cease. In no-energy condition that is the condition when energy remains unmanifested due to absence of matter, all thinking and perception should come to a halt.

It is evident from all above discussion that the law of conservation of energy-matter is only applicable when extremes are excluded. In case of 'no energy and all matter' condition the law should fail because there is no perception left of such a situation. In case of 'all energy and no matter' the law should equally fail for there shall be no one to test it. It is only in the intermediate stages that it would hold well, of course, subject to the condition that matter is found to be not unreal.

Scientists today believe that the Universe began by a big bang and shall end in a big black hole, which finally dissolves in something, or disappears. Whether there should be another repetition of the process or not is a question properly touched by the scientists. Thermodynamics predicts that since the entire Universe shall achieve one constant temperature in due course of time, it shall meet a heat death, in the sense that there shall be no available flow of heat from a system to another. There are many who do not like or support this view of possible heat death of the Universe.

However scientists believe that in a closed thermal system of irreversible changes, the entropy (which can be taken to be measure of unavailable energy or a measure of disorder) shall always increase except when the system is acted upon by external agency, like refrigeration where the cool gives heat to the hot and gets further cooled because the system is acted upon by external power supply from outside (It is however to be noticed that during the cosmic dance of subatomic particles, the energy required to create a pair of particle and antiparticle is exactly the same as that generated when particles annihilate each other and in such a change the entropy should not increase). It follows that if the Universe has to repeat itself from a big bang to the heat death or the disappearing black hole and then again to the concentrated matter to once more blast into galaxies and stars, the Universe shall have to be taken as a closed system acted upon by external power whose nature we do not and possibly can not know being outside our Universe. This power is attributed to God by some religions and some philosophers without of course reference to increasing entropy or the big bang, for these physical terms and their meaning must be unknown to the philosophers of the past. Indian philosophers have said that the world cycle repeats itself after the 'Pralaya' which is a term denoting complete annihilation of the Universe. At the big bang there is enormous difference of temperature and after having passed through stages like formation of galaxies and stars with various kinds of motions and formation of smaller black holes and as speculated by some scientists the ending big black hole consuming the whole Universe, it should come to an end following the heat death. If it is assumed that the cycle is never repeated then it shall have to be concluded that Universe has come into existence only for once out of nothing-ness and shall end into an eternal nothing-ness, an idea hard to digest without convincing proof. If the existence is only for once then it pops out of nothing and in nothing-ness it dissolves. If God is attributed the creation then He is without any work in the previous nothing-ness and in afterwards eternal nothing-ness and thus it points to the end of the active creator after the big bang collapse. Well, if it is proved we accept the transient role of God, which ends with the end of the Universe, but if God is eternal then eternal nothing-ness must be impossibility, for the existence of God presupposes his capacity to have caused some creation. In an entire eternal nothing-ness, where is the God? However some Eastern philosopher call this creator God as Ishwara, and is different from the Absolute Infinite consciousness and this Ishwara is not more real than the creation itself but the Absolute is supposed to be always existent as we shall see later in some more details.

I must not be misunderstood to mean that the Universes to repeat itself is to fulfill the requirement of God who Himself is the outcome of sublime thinking of the learned and the erudite. If science could ever explain Universe without God then the scientific principle on which the existence pivots can be named as God instead of an ontological or theological God. Thus the scientific or epistemological God takes the place of philosophically derived or religiously trusted God. But science, as it stands today, is trying to know the plans of God as explained by Stephen Hawking through his writings. And I feel that even this attempt is only peripheral although it vigorously desires to head speedily towards the center of the periphery. Scientists may try to circumvent the prior and the post nothing-ness of the unrepeated that is only one time Universe, by saying that the time itself began with the big bang and there is nothing like time to mark existence of the prior nothing-ness. Similarly they may say that the time ends with the dissolution of the big black hole in which the Universe should end itself, and there should be nothing like post Universe nothing-ness. But all this is a guess and speculations for want of experimental verifications which scientists of today cannot accomplish. When scientists speak of formation of a black hole of an extinguished star they say that time ends for that black hole only and time shall continue to flow for rest of the Universe. Thus the end of time is in comparison to the time flowing in the rest of the things. Even if the Universe turns into a big black hole the claimed end of the time has to be in comparison to something that continues to exist and who knows there are not many other Universes or an infinite number of them existing beyond intellectual grasp of today's human beings? Just as motion is felt in comparison to a state of rest, the rest is also felt in comparison to some motion. If the time for this Universe should cease, the cease can only be recognized in comparison to a continuing time somewhere in the other Universe or Universes. In the chapter 'Time', we have seen that time is beginning-less and it is end-less too. We have no reason to believe right now that science has reached its peak and that nothing remains to resolve. It is still growing.

There is large difference of kind between a scientific theory and a speculation, although many times speculations have turned into theories much after. A scientific theory is a statement, which in appropriate circumstances and as a result of experiments or when a big logical mistake is detected, is capable of being proved to be wrong. Robert Boyle said that the pressure of a gas is inversely proportional to its volume when the temperature of the gas does not change. Large number of experiments with various gases has always proved his theory. It was always possible and it is always possible that a single exception to the rule could be found to result into collapse of his theory. When this possibility of disproving a theory exists, the theory is known to be scientific. Now suppose a theory is propounded that all men after their deaths go on the planet, which orbits around the star known as Alpha Centauri about 4.5 light years away from the Sun. This theory can not be called scientific because it can not be proved nor disproved. The theory that Universe will collapse one day into a big black hole is a speculation because there is no experiment which can prove this or even disprove this. It does not enjoy the status of a scientific theory like Boyle's law. If we have to believe in some sort of speculation or guess then philosophers definitely have better beliefs at their credits. At least they do not take us from this perceived existence to an absolute nothing.

Chapter 8

Before The Last Chapter

Before readers turn to last chapter, I think it shall be more useful to go through this chapter more carefully. We have to get rid of many worldly ideas when examining the origin of all that exists. Every worldly thing has a beginning and has an end and has duration of time for it to exist within. Everything we perceive must have some dimensions is another idea to be shaken off. Time has been shown in chapter 2 to be beginning-less and the question when it began is meaningless. However the question why it is, is still to be answered. Likewise space being beginning-less and end-less can not be subject of any dimensions or can not have a spot in it where it should begin, but why it is after all, is a question still not fully answered. In fact this 'Why' is more important than 'Where' and 'When' since it goes to the root of any phenomenon observed. As already shown, it is the observer who supplies elements of time and space when he becomes aware of any phenomenon.

If the observed chain of cause and effect is beginning-less, we can not know what was the first event that occurred in this Universe since every effect must have some other event as its cause and this cause in turn is the effect of some other preceding event. This leads us to an infinitely long chain of events when working backwards like this. Now suppose that this chain of cause and effect has a definite beginning, and then it follows applying ordinary logic that the first event in the formation of this Universe must be without a cause or without its preceding event of which it should be the result or effect. Leibniz's law of sufficient reason leads us to think that either this law must be wrong when examining the beginning of the existence, or nothing might have begun or come into existence at all. However what we perceive is to be considered as having come into existence then Leibniz's law of sufficient reason must be wrong or not applicable when we consider beginning of this entire being. Thus we are left with three possibilities to consider-

(1) The chain of causation must be beginnings.

(2) Leibniz's law is not applicable when examining 'Why' of the beginning of the Universe, or it might be wrong altogether.

(3) Nothing might have come into existence at all including of course the observer and the things observed

Point 3 above can be further split into two namely,

(a) Nothing exists at all and nothing is perceived by anyone.

(b) Whatever seems to exist is an illusion, which a non-existent perceiver perceives in absolute nothing, or in perfectly quality-less pure existence.

Point (b) above creates two more fundamental questions namely

(x) Can anything come out of nothing?

(y) Can even illusion come out of nothing?

The reason for coming up of these finer points is very clear. If it is proved that the Universe is not really existent but is simply an illusion, it shall have to be proved that illusion can come out of nothing, if this could not be proved; then there is no existence whatsoever at any level, shall automatically be proved to be wrong. I shall go through these points one by one.

(1) The chain of cause and effect looks universally applicable although we have seen that if the Universe has a beginning, and has the first event of its coming into existence, then this event must be without a cause and applying Leibniz's test of sufficient reason to the first event, it follows that the event must be unreal and logically, it should follow that entire chain of cause and effect bringing us to the today's Universe, must be false and thus the Universe must be unreal. As I have discussed in details the chain of cause and effect (in chapter 6) or what we call causality can not be proved to be existent in the external events that take place one after the other, since our mind has never perceived such a thing like causality howsoever carefully we may have watched a happening. This element is supplied by our minds due to persistent experience of two events one after the other which we conclude to be the cause and effect. Heat is no effect of fire but it is only a sign of fire says Berkeley. We take the fire to be the cause of heat due to our habit or custom says Hume. And so we have no right to say that what we feel in our minds must dictate the external world's behavior. Even if we take the Universe to be series of unconnected events as regards causality, our minds shall always be searching of some chain connecting these events and we shall find some such formula that a particular thing 'B' always happens after another thing 'A' has occurred and we give this observation a name of causality. So we can take this inbuilt causality of our minds, while judging the events, to be the causality which Leibniz's law of sufficient reason conveys. Physicists today say that Universe began at the explosion of the point mass some fifteen billions of years ago, but they could not yet know the cause of this occurrence. In the absence

of the sufficient reason for such a bang, one may conclude that there has been no big bang at all, for this speculated happening is not supported by Leibniz's law. Even the materialistic Universe could not have begun like this. If the role of observer is correctly included in the entire existence then too this beginning is not the beginning of the entire existence. Physicists John Wheeler, Stephen Hawking and philosopher Newton Smith have expressed that the role of observer has to be given due importance in all types of theories regarding the beginning of the Universe.

Now it is clear that there are many logical and even scientific difficulties, which arise if we assume the chain of causation to have a beginning. If the first event is without a cause, the entire Universe proves to be false. This idea of a false Universe can not be so easily digested by the scientists whose purpose is to analyze the phenomenal and material world, and also by most of the Western philosophers, particularly the materialists, realists, pragmatists and neo realists, even if it is proved to them that the Universe is unreal. Thus we have to conclude that the chain of cause and effect is beginning-less. This does not go against Leibniz' law.

(2) Before we discuss the question whether Leibniz's law is wrongs altogether or is not applicable while examining the origin of the Universe, I shall go deeper in the very meaning of cause and effect. Some thinkers hold that the element of causality is inherent in the worldly phenomena while there are as many others who say that causality can not be experienced in the changes observed in the external world and thus the element of causality is supplied by the mind. I say that even the idea that this element is supplied by the mind is wrong. Causality is not perceived in the external world nor is it felt by the mind. What is internally experienced is an association of

ideas of two events which are seen always to occur one after the other and we give it the name causality whereas really even mind does not internally experience the first event that was the cause of next event which we are accustomed to call effect. The rigid association of such two events where the next has always followed the first has induced the idea that the first be called the cause and the next one as the effect. In fact we do not know whether the first event is really the cause of the second. Thus I hold that the idea of cause and effect is totally false, and as far as the perceiving subject is concerned, he can never know that any event can ever be the cause of the other. As I have already indicated, the Universe may be comprised of a chain of unconnected events, where no event is the cause of the other and it is through our experience of the external world, that we develop the idea of cause and effect whereas it, as I said, does not exist at all in this Universe. The idea of cause and effect is just an imagination and nothing else. If this were not true then Berkeley and Kant would not have met so many critics of their theories; howsoever they may look religious, or self-surrendered to the God the Almighty. This discussion leads any reader to think that in fact there is nothing like causality in existence, and what we call an effect of a cause is nothing but a habit or custom of our minds. In this Universe I hold that there is nothing like a cause and there is nothing like what we call effect. It is only a phenomena observed by us who are as real a creation in this Universe as are the external events about which we choose to philosophize. Who gave us the right to judge what the reality is? A rationalist may think that it is the power of reasoning which a man has already got which authorizes us to be the judge of everything that happens before or in the knowledge of us. But can he prove himself to be the sole judge to settle the issue of existence? One can not deny that we are as much a part of creation as are things in the external world and trying to proclaim theory of existence is not in anyway superior, to a leg of a table propounding the theory of existence of remaining three legs or about the flat top of the table.

The Law propounded by the mathematician philosopher Leibniz, that if a sufficient cause can not be found for an existence of thing or a phenomenon, then the thing or the phenomenon is nor real, now must be reoriented in the light of what has been discussed by me, that the causality is nothing but a name given to the always associated events that occur one after the other. Even a single exception to this association is enough to shatter the principle to pieces; is not difficult to appreciate. When the Leibniz's law of sufficient reason is viewed with this new vision, there appears nothing wrong as far as the premise goes. Leibniz must have had the objection about the existence of the Universe without the cause in his mind, but he tried to circumvent it by saying that it is God who created the Universe and He is the first cause of all this existence around us. But I say that this is his first defeat of his law of sufficient reason. He has never proved that the God was the first cause of the Universe, and nobody seemed to object this probably due to the fact that what he proclaimed was in line with the religious beliefs in those days. Rene Descartes and Spinoza also could not free themselves from the effect of dogmas prevalent in those days, and although Descartes is credited for inducting rationalism in the philosophy, none of these could avoid taking resort to God to explain what they could not on pure rational grounds. If there is such an omnipotent entity responsible for the existence of the Universe then all inquiries on rational grounds should have come to a halt after this. But luckily this was not the case. . The philosophical horizons seem to expand much more speedily after what the mathematician philosophers like Descartes and Leibniz gave their valuable part to the treasure of knowledge. No philosopher to come after them, could ever neglect what they said, is sufficient proof of their abilities.

The question whether any law, which has been observed to hold good after the entire creation has come into existence, can explain the origin of the existence is very tricky as far as the logical principles are concerned. Our minds suggest us that there could be some different law holding good at the time of creation or there might have been no law within the comprehension of those beings who came into existence after the creation was done. And it is difficult to decide this for two reasons. 1- That whatever be our verdict that may not dictate the actual happenings in the initial stages of this creation and 2-There might not have been any principle or law when the creation came into being. Physicist Stephen Hawking in his famous book "A Brief History of Time" has expressed that at the time of creation, all laws of physics seem to break. After only a small fraction of a second the laws of motion, the laws about heat transmission and other laws seem to have begun governing the Universe. Considering what I have said that there is really nothing like cause and effect in this Universe, it would seem correct to say that law of sufficient reason is about such pairs of events which show a definite sequence in their occurrence and this, unless an exception should crop up, is applicable always if the Universe is beginning-less. If it had a beginning, there would not be any such preceding event to the first event of creation and the law of sufficient reason, even in the new context of modified idea of causality would certainly break down. When we consider or rather conclude that Universe is beginning-less, then there is no necessity to bring in God as the first cause. If we examine this theory of God, and then according to the law of sufficient reason, we shall have to be prepared to answer the question as to what is the cause of existence of God, for every cause is effect of some previous cause. I think that God has been taken resort to much earlier in the chain of rational thinking by making him the first cause of an assumed beginning-full Universe. In fact the riddle how a beginning-less Universe should come into seeming existence is much more difficult to resolve and it may be here that Gods omnipotence, omniscience and omnipresence may come to our rescue.

3- Anything, we call real by our senses organs and the mind which we find in this vast Universe has a beginning and has an end. Even our Sun, Earth and other beautiful constellations are going to meet their ends in due course of time, although we with our so short a life span need not worry about this future disaster. Now if we say that this principle of a beginning and an end of everything we perceive in this already existing Universe, should apply to this Universe itself which has no beginning as explained above, then we have to conclude that this Universe is not real thing like the ones having a beginning and an end. However no one can ever say that he has experienced nothing in this Universe, that he has seen nothing or heard nothing. The only idea, which can reconcile these two opposite conclusions, is to assume Universe to be an imaginary thing like an image in the mirror, or as a mirage observed in a hot noon.

In the phenomenon of mirage, we see water pool on a hot road due to total internal refraction of light coming from beyond the pool spot, thereby showing trees and other things inverted, and giving impression of a reflected appearance as we see in a real water pool. In mirage there is no water and a fish in this water pool is further height of impossibility. If the entire Universe is a false appearance then philosophizing about its existence is nothing better than a fish in the mirage pool examining existence of water around it. However the question here is whether such an unreal appearance can come out of, or can take birth in absolute nothing-ness or there has to be something, for such a phenomenon to happen as a substratum of it. It is evident here that mirage pool owes its seeming existence to hot air, the things which look reflected due to curving of rays reaching our eyes and the road surface heated up transmitting heat to the touching hot air layer. Thus so many real things could result into illusory water. A mirror and a real object is required to create an image in it. A rope and dim light is required to make it appear as a snake. If we extend this logic we may generalize by saying that all illusions have something real at the root of it. Illusions, like real things can not come out of anything.

If the Universe is an illusion then there is one major difference between water appearing in mirage and Universe appearing as Universe. It is because we have seen real water elsewhere, that we call the appearance in mirage as water, since it shows water like property of reflecting things, while we have never seen any Universe outside this Universe to call it as Universe. It is an appearance, like mirage seen by a man who has never seen water elsewhere at all. So he takes it to be the first water pool and feels it to be the only reality existing. In India it is believed that deer exhausts themselves to death chasing the mirage to quench their thirst and that is why the phenomenon of mirage is commonly known as 'Mrigjal' (Mrig means deer and jal is water in Sanskrit) In any case whether the Universe is an illusion or a real thing, it has to have a real substratum. It may be termed as God or the Absolute or the infinite consciousness, depending on what functions are attributed to it or in what relation it stands to this world of conscious, animate and inanimate objects, forming subject matter of human senses.

The words real and unreal are so very often used in philosophy or science that nobody ever stops to define what is real and what is unreal, and there is a reason behind this usual bypass. What we perceive directly, what we deduce by logic and what is made known to us through a reliable source of information, is normally what a common man takes to be real. All the rest is supposed to be unknown, and if there is no source of its verification it may be taken to be unreal even. A statement entirely against our logic and perception is taken to be unreal; How do we feel at Antarctica may be unknown to one who has never been to that place, but rabbit's horns or son of a barren woman is unreal for us. Water in the mirage is as unreal as an image in a mirror. There are other pairs of words, which are many times used by some thinkers to mean real and unreal, and it is permanent and temporary or eternal and transient. Some Eastern thinkers say that what existed in the past and what exists at present and shall exist in future, is real and all transient existences are false or unreal. This argument has the defect that it presupposes time to be an eternal real thing and the reality of existence is made to depend on this presupposition. Some who have gone deeper than that enquire into the existence of time itself and I have dealt this in chapter 2 in this book. If on the other hand our perception is taken as tool to decide what is real and what is otherwise, the procedure has the inherent defect that it assumes ourselves to be the real thing and our perceptions absolute which is really not the case, as dealt with in chapter 1 and chapter 4. If we take rational thinking and its results as a measure of reality, it can not lead us to it because the hollowness of the chain of cause and effect has been explained in the chapter 6. It would thus appear that we could not accurately define what is real and what is not. Many philosophers say that what we perceive in an external object is all its attributes and qualities, and the real thing in itself is unknown or even can not be known. This is a statement where an unknown thing is marked as real. Thus even unknown things can be real, and a known thing like water in a mirage or an entire movie as seen on the screen is unreal. And even then the thinkers are trying to find the reality in this seeming existence.

But philosopher's and scientist's attempts are not going to be futile in the end of their quests, for the statement that it is hard to define what is real and what is unreal, itself may be false for it has come out of reasoning, a human tool used for centuries together without a common final result in the Western thinking, giving birth to so many 'isms' in the Western philosophy. Similar is the case with Eastern thinking as well but with some difference in the sense that many times intuitions have been seen over riding the logic in the initial stages and then attempts have been made by succeeding thinkers to supplement arguments to what has been already propounded by intuitive power of the seers. Even scientists can not deny that many discoveries have been triggered by intuitions to be later on supported by reasoning. Since scientific theories are most infallible theories, the intuitions must be given its due credit in advancement of knowledge in any branch of learning. Only reason won't do.

I would like to draw reader's attention to one yet interesting aspects of perception of illusions. When we say that the water in the mirage is unreal we unknowingly assume that we are real. Our judgments made by the help of our sense organs are correct, and that the initial sight of mirage is false due to what we see or feel by touch when we approach and reach the place of the scene; but suppose we too are as unreal as the mirage water itself, then what should be concluded of the whole show? It will be an unreal thing perceived by an unreal observer. When the observer and the thing observed both are unreal there remains nothing real in the perception itself. Thus in this case nothing seems to have happened at all. No observer, no scene observed, and thus no observation whatsoever at all. Indian Vedantins say exactly this. Nothing has really been created, nothing is really thus observed, and thus there is no kind of observation at all by anyone in the infinite consciousness which is entirely beyond intellect and qualities and which acts as substratum for this false scene. It is like the un-carved rock mass. All scenes are simply those imagined in the rock mass without a real existence whatsoever.

But when a philosophy becomes the base of any religion, the practical aspects connected to everyday life become far more dominating than the crux of the philosophical principle. No doubt the life styles prescribed by the Vedantins and the Buddhists do not differ much, but the root philosophy is not alike. Abstaining from vices, mental detachment from all that is worldly which is supposed to cause pain and misery, leading thus a pious life, may of course lead to a calm and peaceful world. The Vedantins claim that all worldly things are to be sacrificed in exchange of the eternal supreme bliss which results, they claim, into dissolution of mans limited self into the Infinite Consciousness. Buddhists say that this detachment from worldly things is the only way to conquer pain and miseries of life. A translation by Swami Madhavanand of a verse in Sanskrit from Vivek Chudamani, an original work by Vedantin philosopher Shankasracharya of 7th century may interest readers.

"305- Give up immediately thy identification with egoism, the agent, which is by its nature a modification, is endued with the reflection of the Self, and diverts one from being established in Self–identifying thyself with which thou hast come by this relative existence, full of miseries of birth, decay, and death, though thou art the witness, the Essence of Knowledge and Bliss Absolute"

[Modification - Of nescience, and therefore non-permanent.

Relative existence - Samsara or transmigration]

It has repeatedly and unequivocally said by 'Vedantins' that the Infinite consciousness is beyond grasp of mind. We can not perceive it as we perceive external world or our internal bodily functions. The question remains unanswered as to how a man could realize that he is this Infinite Consciousness and identify himself with it, when it is beyond his mind or intellect which is the only tool with him to apprehend anything that exists. Avadoot Geeta supposedly composed by Dattsatreya purports to describe how one feels when firmly seated in God consciousness. A few verses translated into English by Swami Ashokanand may be found to be interesting to the readers and they may draw their own conclusions on the question of possibility of realizing things beyond intellect.

44 of chapter 1– I am free in the beginning, in the middle, and in the end. I am never bound. This is my sure knowledge that I am naturally spotless and pure.

17 of chapter 3- This (Self) is neither knowable nor the instrument of knowing. It is neither reason nor the one to be reasoned about. It is beyond the reach of words. It is neither mind nor intelligence. How then can I speak this Truth to you? I am the nectar of knowledge, homogeneous Existence, like the sky.

23 of chapter 3- I am pure, very pure beyond reason and of infinite form. I am non-attachment and attachment beyond reason and of

infinite form. I am undivided and divided beyond reason and of infinite form. I am the nectar of knowledge, homogeneous Existence, like the sky.

[Some rhythmic repetitions are due to the nature of the poetic composition in Sanskrit]

However as already pointed out, the Indian philosophy being encircled by religion could admit of almost no new thought for the last thousand years except modifications to the main principles to suit individual tastes. In the West, after fifteenth century, the philosophy developed independent of the religion and I think that is the reason for its free and logical development. However it can not be claimed that it has reached the truth yet.

Chapter 9

The Resulting Model

Before I discuss the new model hypothesizing the origin of the existence, it would be worthwhile to re-capitulate what has been discussed and proposed in all these preceding chapters, so that the readers directly enter into the so far unfathomed truths without having to turn a number of pages backwards to connect themselves to the various references used in this concluding chapter. Although it may be a question open for discussion whether rules of worldly logic are applicable when we inquire into the very origin of the world, the conclusions to be drawn after the world has come into existence, have to be based on logic or justified intuitions. Even a possible conclusion that rules of logic can not apply to the origin of the Universe, is itself based on logic and thus can be taken to be within the realm of logic, since all such thinking is there, after the world has come into being. It is logic only, which places certain things beyond its scope just as when looking in the direction of a house we cannot see it due to an intervening hill and we conclude that house is beyond the hill.

We have seen that matter or the external objects made of matter are perceptible to us only through their qualities which depend on our sense organ arrangement and the structure of our minds, while the reality of the object beyond qualities is never known to us and always remained a mystery.... That there has to be something beyond qualities is inferred from the fact that our sense organs start perception due to this something, for quality is our mind's child and can not be the cause of making us perceive the object.

Time is recognizable only through events which are in the form of change in the position or change in the qualities of objects, but time as such is not perceptible without regard to various events. We become aware of events due to matter being involved in them, and about reality of matter it has been said that it is not known to us. We only perceive the changes in qualities, which we term as event. Special theory of relativity tells us even on basis of qualitative perceptions that an event seen to follow another could have taken place simultaneously with it, and it is the observer's velocity relative to the place of event that decides between simultaneousness and the sequential nature of events. I have established that time causes the events and it can not be the other way round. It is the fourth dimension of which we see only a point, which we call the present. It looks that if someone of us finds himself to be gifted with the power to perceive entire fourth dimension, he may find all past and future events already, as trees by the side of a highway, to be discovered by the rest of us when we reach there or when that events becomes the 'present' for us. One may think this to be pointing to perfect determinism and thus making a place for astrological predictions, but this shall not be dealt here since it is not the subject matter of this small book. These speculations about perfect determinism are yet to be vetted by experiments and perfect reasoning.

The space manifests itself by seeming to contain matter and motions. The warping of space or the pull of gravity is known to us only through events like attraction between bodies. If there were no matter, we could not have got the idea of space to contain it. Space is so to say a relation between two or more objects showing how one is situated in relation to the other. All these are qualities of space, and space by itself, is not known to us except as imagination. Science today tells us that space is inseparable from time, and thus space in itself has to be unknown to the same degree as the time.

Energies and forces can not be experienced or observed without matter and motions of bodies in space. Events that indicate presence of force or energy are perceptible only through observer's capacity of sense organs or their extensions and the mind.

Matter and energy are inter-convertible. In fact the presence of matter or energy has identical effect upon the space-time warping. It is thus clear that out of space, time, matter and energy none can change without affecting the rest of three, and thus as far as perceptions of this Universe are concerned, all the four make one amalgam and manifest themselves as what we recognize as Universe perceptible through our sense organs or through extensions thereof. I can coin a word MEST for this union of these Universe-forming ingredients, that is, matter, energy, space and time.

The perceiving subject that is we are also composed of MEST, and there is an infinite consciousness underlying both the perceiver and the external objects, which is responsible to make the subject feel that external objects do exist. Matter and mind were considered separate by Descartes, and this gave birth to a dualism of its own kind, which now stands annihilated. Without a common substratum of consciousness no interaction of perception is possible, and this substratum is beyond qualities therefore must be identical everywhere and every time, for being at a place or being at a time also are qualities of external objects, and there existence is limited by time and place. As already shown the perceiver himself imposes the qualities. Thus the infinite consciousness must have the power to imagine vast number of different objects on one side and a perceiving subject with limited sense organs and mind functioning on their characteristics of sense organs on the other, and cause what we call perception of the Universe. Where there are no qualities at all, any group of qualities could be imagined. The blank screen of the cinema is capable of showing any type of picture projected on it. However it must not be deduced from above that our minds can change the scene before us. We can not turn a wall into a horse. The reason is that our minds and the object perceived both are imagination of the infinite consciousness and our minds can not change the perceptions, for we are also objects imagined by the infinite consciousness. We are parts of its imagination. If anyone of us could go outside the realm of this great imagination, he could directly perceive this mysterious game of existence, but it is impossible to be out of clutches of imagined world since we are its part.

Now let us turn once more to what science has to say about the existence, and just as no one ever saw or perceived the big bang and still feel that the thing really occurred, I can have another parallel and more convincing guess about the beginning of the Universe.

It is an established fact that matter and energy are inter-convertible and that just as a particle and its antiparticle when brought into contact annihilate each other with release of energy; it is equally true that energy when so directed produces a pair of particle and its antiparticle. In chapter 1 & 4 the subject has been dealt in detail. What we find in the Universe at present is matter and energy both. As physicists say, the Universe is heading towards a condition when there shall be no available energy, and thus no difference in temperature in any two objects. This stage they term as the stage of maximum entropy, and as time passes the Universe is getting near and near this stage. Physicists connect this phenomenon as a factor, which indicates the direction of time; although I do not see any reason to think that the time is in any need for deciding its direction. Time may not be a straight line like entity where its direction should matter and as Newton Smith has suggested time could be a circularly moving entity which has no final direction. There is another way of telling direction of time which scientists call psychological direction, which they define as from memory to no memory. That is the time flows from what we remember to the direction we do not remember. This has also no applicability whatsoever as shall be seen when we enter the middle of this chapter. Memory has many deceptive features and can not be used to decide about any such features of entities like time, space and the chain of causation. So returning to the inter-convertibility of matter and energy and agreeing at the same time with the scientists when they say the entropy is continuously increasing we proceed to guess its effects.

The present stage of the Universe with "some matter and some energy" is a point when we have still available energy and entropy is not yet reached its maximum. The final stage shall be "all matter and no energy". Since scientists and philosophers both believe that Universe follows many physical laws smoothly and invariably, the present stage of Universe must have been the effect of its past stage and extrapolating the phenomenon of increasing matter and decreasing energy, there is nothing wrong to conclude that the initial stage of the Universe must be "all energy and no matter". This has no connection with the over popular theory of the 'Big-Bang' for the big bang is guessed to have occurred when there was already a concentrated matter and wherefrom the huge energy to blow it to pieces had come, is not even guessed yet. I shall try to supplement the missing links even in the big bang and I shall let it have happened but not the way and at the time, guessed by the supporters of this theory.

There is no fallacy in thinking that initially there was "all energy and no matter" stage. The final stage we know to be all matter and no energy. One may say that it is all unavailable energy. But a slight reflection will show that the energy is capacity to do work, and work is defined as the multiplication of force and the distance it pushes the body through. The unit of force itself has mass as one factor and the acceleration as the other. When there shall be no available energy, no work can be done and it will be as good as having no energy at all. If there is no available energy it means that even force of gravity is unavailable. This is possible if all the matter lumps together without even intermolecular or any spaces, which can separate elementary particles from one another, have been consumed, and matter concentrates to incredible density. Now let us visualize the initial condition of "all energy and no matter". When there will be no matter, no bodies, no particles at all it shall be difficult to have concept of space. If there could be some observer like us, who can witness such all energy state without being a party to the affair or without being a part of such a Universe, he would conclude that there is no energy even. Energy needs matter to manifest itself. We need a body heated up to recognize energy of heat, we require some vibrating thing to cause us to know the energy of sound and we need somebody to reflect the light waves to make us feel presence of light energy. Energy without any matter to work upon is no energy or we may name it as 'energy unmanifest'. The outsider would simply perceive nothing except space and passing time and that too on the basis of biological changes in him or by feeling the changes in his mental status as discussed in chapter 3. If such an observer would have continued to be watching the things up to today's state, he would find all matter coming out of nothing, since the unmanifested energy means nothing to him as discussed above. Apparently he should feel that the law that something can not come out of nothing is bluntly violated in the formation of the first matter particle. He would be amazed to find pair of particles being produced, their lumping together in some definite order to preclude immediate annihilation, formation of different elements and compounds and gravity acting to form bigger lumps resulting into galaxies and energy giving them motion in some orderly fashion. A man of science can work out the details and order of such formation in the light of scientific laws that are being obeyed so rigidly by all bodies in this Universe. However such an orderly development can not take place without some plan and more fundamentally a conversion of energy into matter has itself to be willed by someone or something not a part of the whole show. After all why should matter and energy be inter-convertible at all? Why should there exist a pull of gravity whether as a force or as a result of warping of space or due to any other reason yet to be known to the scientists? Why should electron have an electric charge and why a proton has a charge opposite that on an electron?

The planner of these things in the initial stage when there was no matter must not have the material brain like we have, but should have a highest intelligence without the gray matter. We may call it cosmic super intelligence for the sake of proceeding further in this discussion.

The philosophers of the Sankhya school of philosophy of the East had somewhat similar theory of evolution but it grossly differs from the above in as much as they were not aware of the inter-convertibility of the matter and energy and further they assumed one new entity called 'Purusha' which was independent of the initial conditions which they called 'Prakrati'. Sankhyas claimed that elements such as mind, intelligence and the feeling "I am" which is known as ego or 'Ahankara' can not be generated out of 'Prakrati', which they said was basis of the physical objects. They attributed existence of such elements to 'Purusha'. According to them when 'Prakrati' somehow came into contact or even in the proximity of the 'Purusha', it began evolution for the sake and enjoyment of 'Purusha' who is supposed to be an independent witness having all knowledge but no direct interest in the evolution. This theory of 'Prakrati' and 'Purusha' is also only remotely suggestive of the Darwin's theory of evolution. The former is teleological while the latter is more scientific.

Another point which invoked criticism on a large scale was their postulate that there were more than one 'Purushas'. It seems the multiplicity of 'Purushas' arose out of confusion that on one hand they claimed the 'Purusha' to be absolute uninterested being while on other hand they tried to reconcile the effect of many people who perceived or enjoyed the evolution out of a single 'Prakrati'. When they say that 'Prakrati' was infinite it can not be understood how 'Purusha' came into its contact, for to come into contact one has to be out of contact a moment before and this was not possible for an infinite and all pervading entity to keep some other entity out of its touch. However this theory of 'Prakrati' and 'Purusha' at least takes the knower into consideration and suggests that nothing would have been created if there were no contact or proximity of 'Purusha'. Scientists like Stephen Hawking and John Wheeler have admitted the role of the knower or the intelligent being in the existence of the Universe. 'What good is a Universe without somebody around to look at it' was said by Robert Dicke; and John Wheeler quoted him in an interview with Cosmic Search. Wheeler also added 'that, to be sure, was an old idea going back not only to the Bishop Berkeley of time of Newton, but all the way back to Parmenides, the precursor of Socrates and Plato'.

It is of course to be remembered here that all matter or all energy conditions depend upon the qualitative and quantitative knowledge of phenomena in this Universe and as discussed in chapter 4, our knowledge does not tell the real nature of things. What we know or perceive are the qualities of objects only and not the reality of the thing in itself. The thing in itself is ever a mystery for us. This logic is equally applicable to other man who is the knower of any object, which we perceive. The other man is just like any other object whose qualities only we are able to know. The real thing in itself of that man is also ever hidden from us. Philosophers of the West generally take the man as separate entity from the perceived objects and they assume that only such things are true which man judges to be so. They separate the knower from the known so much that this dualism has become problem to many philosophers who try this way or that way to reconcile the two entities that is matter and the mind or what is material and what is spiritual. Readers may choose to go through chapter 4 once more if that is felt necessary to refresh their memory on these issues.

It is evident that the condition of all energy and no matter would not have any witness like us since we are made of matter and energy both. If we were to go back to this condition by reversal of time element we would simply dissolve into the vast calm emptiness of unmanifested energy only. If causality is a real thing in this Universe then we are the effect of that 'all energy condition' and one would be wishfully tempted to feel that he was always there in some form and he shall always be there. Eastern philosophers support this view but in much modified form, which I shall come to later on. After all the thought that we are not here only for ninety or hundred years out of infinite looking time, and that we are also eternal as much as the time itself, is definitely soothing and encouraging, and exciting also to feel that we were with Plato, Aristotle, Jesus, St Augustine, Buddha, Shankaracharya, Newton, Spinoza or Einstein in that all energy condition.

It would equally be impossible for an observer like us to perceive the 'all matter and no energy condition' for it would need reflected light to see a substance, it would require vibrations to make us hear the sound, it would need diffusion of small particles in some medium to cause a smell and so on, but there would be none of these. In absence of a thing like gravity, which is a force-requiring acceleration to define, even weight or the existence of mass can not be felt. For feeling something it takes energy. Thus "all matter condition" is equally imperceptible to witnesses like us. But when it becomes clear that this condition is the ultimate for everything that exists or has existed in the known past it would be a heartening idea that we may, or we should find ourselves in closest association with Leibniz, Descartes, Hegel, Kant, Beethoven, and Nicholas Paganini in that concentrated lumped matter. After all why should all of us exist only for a few numbers of years in this Universe? Are we really satisfied with this transient existence of us? If our existence is so momentary in this Universe, then why should we have chosen this plight? Were there not better alternatives? Leave aside what scientists say; leave aside what Western philosophers say. If we rise to our inner voice do we not feel that our being born here in this Universe carries a big meaning? May be that we live as energy or as matter, but we in our hearts feel better if we are told that we are eternal. In the Indian highly esteemed scripture called Geeta, Lord Krishna tells Arjuna that we were always there and that we all shall always be there and thus it is futile to be aggrieved by the result of the famous battle of Mahabharata which was to be fought soon thereafter (verse 12 Chapter 2)

We started from a stage when there was no matter and we head towards a condition when there would be no energy. But still we do exist in this concentrated form of matter. And if the cycle of being is to repeat once more or any number of times then we exist eternally. And it is here that I say a big bang is possible. And if that is the reality then the present configuration of matter in this Universe may be justified up to the point when or if the Big-Bang theory is refuted by some still superior speculation. Be it whatever may; the crux of the question before us is whether these speculations can be taken to be real? Or since these speculations come from a man who is a part of this Universe, they should be out right rejected as not being real or these should be assigned a status of real and not real at the same time-space. This reminds of a very interesting deliberation in 'Rigweda' of Eastern Philosophy, which as translated by Max Mueller and runs thus-

"There was then what is not what is, there was no sky, nor the heaven which is beyond. What covered? Where was it and in whose shelter? Was the water the deep abyss (in which it lay)

There was no death; hence was there nothing immortal. There was no light (distinction) between night and day. That one breathed by itself without breath, other than it there has been nothing.

Darkness there was, in the beginning all this was a sea without light; the germ that lay covered by the husk, that one was born by the power of heat (Tapas).

Love overcame it in the beginning, which was the seed springing from mind, poets having searched in their heart found by wisdom the bond of what is in what is not.

Their ray which was stretched across was it below or was it above? There were seed bearers, there were powers, and self-power below and well above.

Who then knows, who has declared it here, from whence was born this creation? The gods came later than this creation, who then knows whence it arose?

He from whom this creation arose, whether he made it or did not make it, the highest seer in the highest heaven, he forsooth knows, or even he does not know?"

The interesting part of this is that this finds its place in 'Rigweda' which is the oldest scripture of Indian philosophy and what followed thence is simply additions of foot notes or modifications of the main idea to suit the convictions of thinkers that came later. This is the first hymn in the Indian philosophy, which must have triggered the acute debates culminating in the philosophy of Acharya Shankaracharya of seventh century A.D.

Both all energy condition and all matter condition presuppose existence of space and time which are elements supplied by the mind of the knower as clarified in chapter 7. When in the all energy condition there was no mind like ours and if time and space were existent, these must have been supplied by the super cosmic intelligence and the thing continues till the act of supplementing these elements have been taken over by men or the knowing entities so to say and thus we become part of the cosmic knower it appears. However there seems to be one difference between the two. The cosmic knower is the knower and the known both for there was nothing else; but we appear to be knower only always wondering about the real nature of the things perceived. The Big-Bang theory which has the General theory of relativity as its base, postulates that the time and space both began with the big bang. So the question what was before that is meaningless. But this speculation gets diluted in reliability when it is said by the physicists that the big bang finally ends in a big crunch and then the cycle may repeat over again. By saying this, the thought that time could be there before the big bang, gets ground and the subject is thrown open for discussion again. If such a repetition is a reality, how many of them have occurred already before the present Universe came into the present shape, shall be another point which finds no satisfactory answer at all.

I have no hesitation in admitting that the postulates of Indian philosophy which were difficult to understand being in verse form, which was considered to be the briefest form of narration, and to commemorate, were easy to understand after I studied the Western Philosophy which I consider to be most scientific and analytical in nature. If I had not studied the Materialism, Organism and Organicism, Neo realism and objective realism at length, I would not have had an insight into what 'Charwak' had said in those days which was intensely similar to the materialism and pragmatism of these days. The reason is obvious. The Indian Philosophers did not go much in the argumentative aspect of their theories, but had simply narrated the result of their deepest thinking. I should venture to say that many arguments which should have been given by the ancient philosophers appear to be supplemented by the Western philosophers in their deepest investigations of the similar philosophical problems. The Western philosophers were argumentatively more articulate than their Eastern counterparts, is truth, which can not be refuted. Science and scientific thoughts go together. The intuitive solutions of problems which vexed our minds since ages were offered by Eastern philosophers, and it remained the job of intelligentsia to logically prove what the great thinkers accepted as the truth of existence by their intuitive insight.

Whatever are the facts of history so remote, it has to be admitted that the search for the truth was a continuous process in East and West with equal vigor. And why should it not be? In the East some of the philosophical findings were so strongly considered to have reached the truth that they became way of life in the form of religions. Buddhism, Jainism and Hinduism are examples of this fact. However this had the disadvantage of philosophy having been encircled by the religious austerities, which narrowed the latitude of further additions or modifications to the old established principles of philosophy. In the West, after fifteenth century, philosophy looks to have come out of religious domination mostly due to scientific discoveries and strong faith in mathematical truths, and this continued to keep open newer avenues for philosophical thinking. At present the religion and philosophy look going ahead unhindered by one another in the West. However this not being the subject of this book let us turn again to scientific and philosophical aspect of this perceptible Universe around us.

Universe was initially in all energy and no matter form has been asserted on the presumption that man is the knower and that the external world, that is the objects outside our minds exist. There are many thinkers like Leibniz, Locke, Bertrand Russell and many others who believed that external world really exists. But there is an equal number or even more who believed that there is nothing beyond the mind and that the mind is altogether different entity from material objects and thus it is impossible for external objects to occupy the mind by there presence. The minds working stuff are ideas and thus only ideas can enter the mind. Since the objects if there be are perceived in the mind there has to be someone to convert the perceptions into ideas and then this someone must have the power to charge our minds with those ideas resulting in what is called perceptions or experiences. Berkeley declared that this someone must be the God Himself. Kant and many others more or less follow same line of thinking with of course some modifications, characteristic of their thinking and intuitions.

It is thus now very important to know whether the external world really exists or it is some sort of illusion persistent enough to make us believe so intensely that it does. If it is proved that the material world does not exist, then we shall have to believe that only mind is the existent thing. However the difficult question would then arise as to 'Whose mind' really exists, for then my mind may be just an external object for anyone else's mind. It is agreed that mind is not a material object, but its presence in some other person is felt by me through action of his body only and there is nothing much wrong to say that other person is a material object having activities like speaking, walking, eating and the like, just as any living creature has and all such creatures are taken as objects of the external world. Darwin believed that the mind has come out of the material known as brain and its birth was due to the desire to exist and protect the life. The Eastern way of telling the same thing is that mind comes into being when the consciousness unites with the life-force or what is called in Sanskrit as 'Prana'. This life force has an in-built constant urge to continue to exist. Even animals and insects with very low intelligence change their course of movement when some danger to their life is detected by them on their way, and this act must be the result of their thinking howsoever inarticulate that may be, but it proves the existence of mind in them. In a just born young one of an animal the mind is not so developed as to cause a protective action against a danger, and we may conclude that the consciousness and the life force both are very weak in that just born condition. Thus only physical actions of other beings suggest existence of mind in them, for we are unable to directly perceive the others mind. These actions may be speech, writing, walking, running and what not.

In the chapter of matter, I have explained that we are able to perceive only qualities of objects and not the object in itself. The apparent qualities are dependent on the arrangement of our sense organs and frame up of our minds, which makes the judgment about existence of an object as recognized by its particular set of qualities. The qualitative appearance of an object solely depends on the perceiving subject or the observer. Different observers may throw a different set of qualities on the same object and we can never know how the color, which we call red, appears to other man. He will of course call it red because he has been taught to do so from his childhood, but how it really looks is a mystery for everyone except him. Einstein in a conversation with Rabindranath Tagore has said "The same uncertainty will always be there about everything fundamental in our experience, in our reaction to art, whether in Europe or in Asia. Even the red flower I see before me on your table may not be the same to you and me"

We can define the external object now as the thing in itself, in addition to the appearing qualities or say A+ set of qualities for a particular object. The other object can be expressed as B+ set of qualities of that object. The third can be expressed as C+ set of qualities of that third object and so on for various objects and so are the things in themselves which we can never perceive. A deeper reflection will show that since A, B, C are all without any qualities, it is impossible to tell one from other, for it is only set of qualities by which we can differentiate between objects. Shape, size, color, temperature, weight, toughness and sound taste are all qualities beyond which the A, B, C exist, but we can not differentiate between A, B and C and so there is no point in calling them by different symbols. They are and have to be identical and we may name it as X only and do away with different symbols. Thus A, B, C etc are all identical with X and this can equally be made applicable to other animals and human beings, which are external objects to a particular observer. Logic leads us to the inescapable conclusion that the thing in itself of all objects in this Universe is only one that is X. It would seem that on a big canvas of X there are many paintings with different qualities and this is what we call this world or Universe. The thing in itself of Sun, Moon, stars and all things is one and that is X. The qualities are not absolute and can not be called as real whereas the thing in itself that is X is one and the same for all, including the object, human beings and the observer himself. A doubt may be raised here that when this X is not subject of perception, why we should take this to be existent. This doubt can be removed when we once more look to what have been discussed in the chapter of matter and that of knowledge. Almost all philosophers of the West and East have agreed that what we perceive is only qualities and not the thing in itself, thereby establishing that thing by itself does exist and it is beyond qualities. Secondly if we go into deep as to why should the perception of qualities begin at all, it shall have to be said that qualities themselves do not begin the appearance of the object, for the set of qualities is thrown upon the object by the observer himself. The qualities do not exist at all in the object itself. So there has to be something beyond qualities, which excited our mind and sense organs, and as a result we throw a set of qualities on the object, and this something is the reality of the object or the thing in itself. I hold that the reality or the thing in itself of the object and that in the observer must be exactly alike and must be able to vibrate so to say in the same tune to cause the perception. This also removes the difficulties in the philosophy of Descartes in which he considers the mind and the matter as altogether different elements and philosophers went on suggesting theories by which the two altogether different entities could interact to cause the perception. I say that perception is due to the exactly same entity common to the object and the observer and there is no heterogeneity in the process. The X is a homogeneous element. It may be imagined to be like a net spread all over in all the objects and all the observers and its vibrations (imagined) induces the observer to throw a set of qualities on a particular scene and thus the perception is triggered.

Another doubt may be raised here as to why should objects look different from each other when there is the same X everywhere, or put in other way why should observer throw different sets of qualities at the places where different objects exist. This question was left unanswered in the chapter on matter. Now I propose to reply this. The fact that one object looks different from other is itself a quality, and not a separate element from the bunch of qualities. In the bundle of qualities we have simply to add one more quality of capacity to look in a particular way. The horse looking as horse and a car looking as car is due to the fact that there is one more quality to be added to the bunch of qualities of a horse that it should look like that. Thus the different objects have got one additional quality of their appearance in a particular way, and this quality is taken with all other qualities form the total qualities of that object. It has nothing to do with the reality X in all such objects, for the reality or thing in itself is by definition beyond all qualities. And if entire Universe is taken as one object of perception having different colors at different places, different toughness at different places, different sounds emitted from different regions, then all these can be lumped together as qualities and the X is still there beyond all qualities. Thus the doubt about different objects gets dropped when the whole Universe is considered as object of our perception.

Whatever has been said above has got far-reaching implications. As already pointed out, the mind is the creation of union of the consciousness and the desire to continue to exist and not to perish. It is one of the qualities of the conscious living beings to have an urge for continued existence and to resist its destruction. It is a fact that Descartes considered mind and the matter to be of different nature so much that one could not have affected the other. But here we see that the underlying reality of all the objects is able to induce the living being to begin perception. If only live things are able to interact, then the underlying reality or the X of this Universe must be of nature of consciousness, because then only it must be able to have impact on the consciousness of the living beings. Then it has to be concluded that the difference between a living and a non-living object is that, while consciousness is common to both, the non living have no life force or the desire not to perish and thus they have no mind for mind is product of consciousness and the life force. Thus the Universal consciousness gets its living and non-living forms only due to presence and absence of the life force or the desire to continue in existence. I strongly hold that if objects did not have a conscious substratum, our conscious minds would never have perceived them. I can expect many objections to this theory that consciousness when combines with life force or desire to continue the existence, gives birth to what we call mind. Many animals have their minds most under developed. They are not aware of the possible forces that may destroy their bodies. It is in the latter stages of evolution that minds grow enough to be conscious of the thoughts in the minds. Memory too develops in steps. A dog has a better memory than that of a pig. Man is a much-developed creature in the long journey of evolution and has developed minds reaching far out of their mere main purpose of self-protection.

An objection to this theory about generation of faculty which we call mind, would be that how it is that all things have not generated minds. Why should stones not have minds?

This question is like asking why some of the substances are liquids some other solids and the remaining gases at say 25 degrees Celsius. The only reply to this question is that the variation of states of matter depends on our mode of perceptions, that is, at a particular temperature or time or space. Similarly we can say that we perceive Universe in a particular mode, and that is in present tense always and we find that all have not generated minds. They may generate their minds in a distant future or they might have got minds in remote past. It has always been considered irrelevant to ask why a thing is as it is and why not otherwise, because if it were otherwise same question could have been raised as to why the thing is not as it is today.

This may have put some life in the physicist's mechanical world of dead objects orbiting around here and there without any known purpose, and then forming black holes here and there and then the black holes themselves evaporating with no reason and purpose. However as already pointed out, physicists have now begun to recognize the role of intelligent life or the conscious beings having able minds in these otherwise bewildering speculations about the origin of the Universe or the time or the space. It may be due to realization that consciousness has got something to do with this huge appearance of this phenomenal world.

Space as discussed in the chapter 3 is a creation of human mind. Space is an idea about relative situation about objects. Curvature of space is another idea about it. Warping of space due to presence of objects and energies is still another idea proposed to suit the observed gravitational pull among objects. The ideas are simply tailored to fit the observation and as such they must not be allowed to mean that they are the real cause of the gravitational pull, for in years to come there might crop up still a new idea to explain the age old observation. As explained, when an object is perceived, it is the mind which gives it dimensions like space, time and even so called cause and effect. When all qualities of an object are removed what remains is not identifiable as the object but as proved above, the Universal consciousness continues to be there. This further goes to prove that this Universal conscious pervades all space, for it is space which remains when all qualities of the object are taken away. Thus whether there is any object or not, the consciousness always exists.

But this is not all. Why consciousness looks to be objects at places, and why it is simply consciousness without objects at others, has to be explained yet. When there is no observer and no object the consciousness still pervades. The question can be put in a different way now. Why should the Universal consciousness look to have become observer at a place and the observed object at the other? In fact the observer and the observed objects or say the knower and the known must have had their existence recognizable as such due to the qualities they have superimposed on the Universal consciousness which is the substratum of all existence. The qualities come into picture when an observer is there. Without an observer there are no qualities at all. Science is the study of qualities and the phenomena that are observed, but for this to happen, scientists or observers are necessary. Without them the qualities do not even exist. Study of science assumes the existence of the observer and the observed things. But the question is as to why the Universal consciousness should assume the forms of an object and that of an observer at all? Some philosophers without investigating the nature and existence of this Universal consciousness directly bring in the Creator to explain the world's existence and the things become very simple. The Creator is supposed to need no material to make the world out of. He just created and we live in it. But if due importance is given to the existence of the Universal consciousness; the theory of creation by a Creator shall have to be modified. What I hold is that a different meaning shall have to be given to the word Creator now. Let us call the agency which created knower and the known with all qualities out of the Universal consciousness, as the Creator for it is almost equally tough job to create out of this consciousness what is not existent at all in it and which we call qualities. Since the qualities depend on the observer, this Creator has to create both of them simultaneously. Now if the law that nothing which is non existent in the cause can come out as effect is true, then how can qualities, which are non existent in the consciousness, ever be created? There is no other material whatsoever available with our newly defined Creator. The bunch of qualities does not exist without the observer.

Philosophers generally take the position of a judge really existent, and then they proceed to judge the reality or otherwise of the external world. Some of them look oblivious of the logical conclusion that the objects and the perceiving minds are either both real or both unreal. In fact if the object of observation is unreal so also must be the observer. In a picture of a car with a driver inside we can not take the driver to be more real than the car. Both have got the common canvas. The universal consciousness is the canvas on which the objects and the observer make their appearance with equal reality or falsehood. The Creator did not make one creation more real than the other. However the other question rigidly connected with it is that when the consciousness does not have any qualities, how these qualities could be generated out of no qualities. The Universal consciousness can not itself turn into qualities or have tagged on the qualities from nothingness. However if we go in much deeper it will be evident that it is the observer who is responsible for appearance of qualities to be superimposed on the consciousness, but as already been discussed, the observer himself is a bundle of qualities for another observer. It shall have to be unavoidably concluded that the observer and the qualities of various objects observed, have a simultaneous birth and we can not say which came first. If there is an observer, there are qualities in the observed objects. If he is not, nothing ever existed as an external world. It shall have to be concluded thus that if the Creator according to my definition of the word can create qualities out of no qualities, then the creation is real. If he can not then the creation is false, and its appearance is nothing but a mirage from both the observer and the object point of view. That is neither the observer is real nor the object before him. All scientific truths owe their truthfulness to the reality or otherwise to the scientists themselves. And coming down to ordinary world of our lifelong acquaintance, the scientific discoveries are real and reliable since we consider ourselves to be real and reliable. Science is study of observed qualities and phenomena and as far as they are real according to the discussions above, they are of immense value to the worldly knowledge and utility to the humanity at large.

It shall have to be remembered here that if the principle that nothing can be created out of nothing is the truth then the entire creation and hence the Creator according to how I defined him are unreal, and what should appear before us is simply a hallucination or an imaginary thing. Of course the base upon which this hallucination must be erected is the Universal consciousness which is real since it is free from the observer, and the qualities he throws upon the object according to his feeble sense organs and the mind, which owes its existence to the consciousness and the desire to live. Even for a mirage to exist there has to be the hot air and the land and the light. There can not appear an imaginary object without something real to serve as a base for its appearance. This is what we observe in our everyday experience. An image in the mirror can not be without a real object somewhere and so is the case with mirage. A man suffering from diplopia can not see two Moons unless one of them is there. If we can extend the analogy to the qualitative appearance of the world, the Universal consciousness must serve as the base for the unreal qualitative Universe. The reality or the unreality of this qualitative Universe depends upon whether the Creator according to my definition of the word can create something out of nothing or not. We do not know what he should have done.

Now imagine this all-pervading Universal consciousness without any qualities, to be like a big rock mass. This is for the sake of imagination only. It has really nothing to do with a rock, which has so many qualities. Now in a rock mass there could be infinite number of un-carved figures, but they are not visible since they are not actually carved out. We may call these figures as statues unmanifested and they can be infinite in number. Thus in this consciousness there could be infinite number of unmanifested Universes along with pairs of observers and the objects.

Now it is here that readers may find it necessary to refer once more to chapter 1 wherein the dreams had been dealt with. As soon as the dream begins, the perceiving subject and the dream objects make a simultaneous appearance. We have never heard of a dream in which the perceiving subject was not there, nor do we have any record of a dream where the dream objects were not there. If I lie in my comfortable bed and dream that a big Lion is chasing me, neither I of the dream nor the lion of the dream is more real than each other and nor the entire scene in the dream has a simultaneous origin. And so long as the dream continues, the 'I' of the dream and the Lion of the dream appear to me as a natural part of my dreamland. This 'I' of the dreamland is different from the sleeping 'I' is as clear as anything, for whereas 'I' of the wakeful state is in his bed the 'I' of the dream is running as fast as his legs can take him to save himself from the dreamland Lion.

In dreams; the dream space, the dream time, the dream objects and the dreaming subject have their existence in the consciousness of the sleeping man. It is because of this common substratum of the mans consciousness that the dream perceiver perceives the dream world which consists of the dream space, dream time, dream matter and dream movement or the dream energy. If this substratum were not common there would have been no perception at all. If A dreams to be with B at a particular dream place and time, B does not dream to be with A at the same time because the consciousness substratum of A's dreamland and that of B's dreamland are altogether different. The dreamland of A is A's creation in his consciousness and not in B's. Similarly I hold that the object's space, time and motion in this world or Universe and the perceiving person are in one consciousness which causes the perception. There could be many other Universes in these consciousness substrata but these possible Universes can never be perceived by us, since we are part of this Universe (unless some supernatural powers makes this possible). Now to accommodate many or infinite number of Universes I would rename this consciousness as infinite consciousness and I shall be using this name hereafter.

The Common consciousness of the waking state and that in the dreams is the same. During the sleep the mind begins functioning and superimposes on the substratum of infinite consciousness, its ideas, which could be in the form of commemorated things, emotions, liking and fears and begins creation of the dreamland. The 'I' of the dream does not know when the dream began and thinks it to be eternal. All this much resembles the creation of this Universe, which we perceive and think to be true and real in waking state. The Creator superimposes his ideas on the great consciousness and makes the Universe we find around us, and we do not know its beginning and its end. It would seem here that I agree with Berkeley and Hegel so far as Universe is looked upon as materialization of Creator's ideas, but the main difference is that I take the infinite consciousness to be existing already and the Creator superimposes his ideas on this substratum. The process of creation may look real, but since something can not be created out of nothing, the creation in it is unreal. It is real that dreams are seen, but it is equally real that they are unreal, and what we see in the dreams is all false. So is the case with the Universe. It looks eternal like dreams. It looks solid and real and obeying some laws just like dreams look to us while dreaming as very real and obeying some laws of the dreamland.

Now if I am asked to define the Creator more exactly, I would say that the infinite consciousness is the Creator, the Substratum of all seeming creation, and the Creator which I defined earlier is some subordinate entity which owes its existence to the Infinite Consciousness or the Creator and is as real as the creation. If the creation is unreal the subordinate Creator is also unreal. When the driver is driving the car, both the driver and the car are equally real. But if a painting is seen of a car and a driver inside, we can not validly contend that the driver is more real than the car, both are equally unreal, but what is real is the canvass on which the portrait is drawn. So is the infinite consciousness is the substratum or the canvass on which the all qualities and phenomena of this Universe are laid and spread? The role of the Creator as I have defined that word lies in simultaneous creation of the perceiver and the things perceived out of the all-pervading substratum of infinite consciousness. The question of creation of dream objects is exactly similar. While dream in the dream world is as real as our waking state world is during our wakeful state. The dream world is unreal for a waking observer, and the waking world is equally unreal for a dreaming observer. This reminds me of the argument that Descartes advanced in this regard. According to him we dream and accept content of our dreams as true but once we are awake the dream is proved to be false, and if the dream state is compared to the waking state we are more often than not placed in a dilemma. Can anything determine for us whether that which we are experiencing at the moment is a part of a dream world or an element of reality? How can we distinguish definitely and distinctly between the dream state and the waking state? Is a dream false only because it does not resemble the waking state? Why would it be wrong to consider the waking state as untrue because it does not coincide with one's dream world? Which of the two is the real world? It is perfectly possible that everything we observe and experience in our waking moments may be false, that our real world may be nothing more than a trick of our imagination, that it may have no existence outside the limits of our mental capacity. Thus Descartes doubts the so-called real world and he maintains that it is this doubt that proves his existence. What is there that can be thought true? Perhaps only that nothing in the world is certain, he adds. So much about dreams as Descartes thought.

Now the question arises as to why should the dream begin at all. The readers may recall what has been discussed in chapter 'Cause and Effect'. In fact the phenomenal Universe does not have any causality between events. Our minds also do not create causality. We simply call the preceding event as the cause and the next event as the effect if the pair of events is always seen to occur. The first scene or the event in our dream by the definition of the word first does not have any preceding event at all, and as such there exists no such pair of events to be termed as cause and the effect. Thus we have to agree that the question why dream began is no question at all for the word 'why' assumes that there has to be a preceding event which actually is not. However once the dream is on the perceiver of the dream considers the dream to be beginning-less or eternal. He does not know and can not know the beginning point of his dream. The creation for him is real and continuing while the dream is on. But there is a reply for the question as to who is the Creator of the dreamland. It is the man who has gone to sleep. It is he who creates another observer in the dream whom he names as or considers being 'I', and it is he, who creates the dream world to be perceived by the dreaming 'I' simultaneously. He is the Creator of the dreamland out of his own consciousness substratum. However the 'I' of the dream may advance many theories like a big bang according to his dream logic for the configuration of matter and energies of his dream Universe. That may not have any reality in it when we wake up out of this wakeful state dream, which appears to continue forever. In India there are some who think that God gave us the power to dream with the specific purpose to make us realize the way this Universe has come into existence. However it is hard to agree to this.

This all pervading infinite consciousness appears to our minds to have taken the shape of this Universe and this appearance can be compared with the appearance of mirage where there is not a drop of water. All qualities appear to exist in a consciousness without any qualities at all. Thus this appearance of Universe should only last as long as our minds exist. If the mind dissolves into the infinite consciousness, the world vanishes. All scientific discoveries and advancements are regarding the qualities and phenomena observed in this Universe by our minds and thus have to depend for their existence on our minds only. Time, space, energies, matter and its qualities and all studies about these entities depend for their truths or otherwise and even their existence, on our minds only. If however we feel that the principle that nothing can be created out of nothing should also apply to the beginning of this Universe of qualities then we can not escape from the conclusion that nothing has really come into existence and hence nothing shall ever go out of existence. But if we assume that the principle should not apply to the initiation of the existence, then the creation of our minds and hence the creation of the space, time, matter, energies and the entire qualitative Universe has come out of the infinite consciousness, and assuming that all creation should have a Creator to cause it, then the entire Universe owes its existence to the Creator. Science must ultimately strike a zenith, which shall maneuver it to this truth. Even the physicists are heading towards almost uncertainty in all respects as to popping of new particles from nothing without knowing reason. The entire discussion can be summed up as follows.-

Space and time are not different from each other as they used to be to the world of eighteenth century. Also matter and energy are one due to there inter-convertibility by equation $E=MC^2$. Space warps due to presence of matter or energy, Thus nature of space-time is also dependent on matter or energy. There is nothing wrong in concluding that space time, matter and energy are so very rigidly and inseparably connected that it is not possible to keep anyone of them the same or unaltered, with any of the other three varying in any manner, as physicists now believe the ultimate curvature of space time due to presence of concentrated matter resulting from death of a star of bigger size leads to singularity as is logically deduced from Einstein's general theory of relativity. Philosophical discussion in this book leads us to the conclusion that there is one substratum without any qualities as support of all perceptible objects in this Universe with all qualities known to us or any other perceiving animal. Since this substratum which is named as infinite consciousness in this chapter is without any qualities, including the size or shape or any other property of matter known so far, it could be a point or a non-point or it can be neither a point or a non-point since calling it point imposes a quality of not being non-point, and calling it a non point imposes a quality of being not a point. Thus although worldly logic says that anything can not be an A and a non A in the same time space, it has been clearly brought about in chapter 7 that logic when made to be applicable to the origin or the end of the Universe has to stand amended to the effect that law of contradiction and that of exclude middle are not applicable to these end conditions of the Universe. The singularity as imagined by the physicists resembles the infinite consciousness in as much as its location in space can not be ascertained, or as physicists say the space and time were non-existent before the big bang occurred. The infinite consciousness is also beyond any qualities like being bound by space or time. The two also look alike in the sense that both of them have no known origin. It is not known when and how the concentrated matter came into existence. Similarly the when and why of the infinite consciousness is also unknown. The point of difference between the two is that it seems that the point matter came into existence out of nothing whereas the infinite consciousness is eternal. The reason for all this creation can not be derived from the process of creation as speculated by the physicists in the form of the big bang. If Leibniz's law of sufficient reason is brought into play here, the Universe proves to be nonexistent for want of sufficient reason for its existence. The infinite consciousness is much more pervading than the singularity assumed by the physicists. In fact it is this infinite consciousness that is responsible to make our minds know the external world as explained in the previous chapter. The mind and matter duality posed by the famous mathematician philosopher Descartes stands nullified.

However why this world appearance should begin at all is a question yet to be answered. Since the reason is exactly not known, as stated by Leibniz through his law of sufficient reason, we assume that the Universe did not and does not exist at all. But then what is it that we perceive? Well, we are just a part of this Universe as trees and stars are. If both the perceiver and the thing perceived are non-existent, what is felt to exist including the perceivers the perceptions and the thing perceived must all be an illusion. Since no illusion can be without some real existence at the back of it, it must be the infinite consciousness as the substratum, which appears to us as the Universe. In fact it is one part of the illusion that is ourselves which perceives the other part of the same illusion as the Universe. In a picture on the wall the boatman cannot be more real than the boat or the water upon which the boat is rowed as shown in the picture. The way with the Western philosophers is that they think themselves to be the real absolute judge as if outside the Universe, and begin philosophizing about the perceived world. They are oblivious of the fact that if the world is real only then they are real. If it is unreal, then they too are so.

Whether the world is real or unreal can not be inferred from our perceptions alone, since these perceptions of the external world take birth in our property or quality to perceive. If we did not have this quality, nothing would have been seen, heard or tasted or smelled and we would have been devoid of any idea about the world.

The infinite consciousness is without any qualities. The words infinite and the consciousness must not be taken as adjective and a noun derived from an adjective respectively, for in that case one may take it to have at least those qualities suggested by these adjectives. To express any idea we have to use some word, and origin of all words and languages lies in necessity to convey something about worldly things. Thus all words convey something worldly or some thought based on worldly things. And we are forced to use such words to express a thing without names, forms and qualities, and each adjective limits the adjuncts of an object, as when we call something black it excludes all that is non black. If we say that the ball is spherical it limits it by conveying that it is not square or conical or cylindrical and so on. The infinite consciousness denotes some such entity without qualities, but due to which the world of qualities appear before us, who are themselves a part of the world of qualities having quality of feeling that they perceive the rest of the world. In fact the infinite consciousness is indescribable since we only describe qualities. Thus if someone asks how the infinite consciousness is? The seer has to be silent; thereby communicating its indescribability, for any word uttered limits the infinite by some positive or negative adjunct. If the Universe is really real the reason for its coming into existence must be known, but since the position is indicative of the opposite, the Universe must be non-existent at least in the form of what is perceived by us, and as non existent is perceived, the entire perception must be an illusion only. The question as to why the quality-less infinite consciousness becomes a world of qualities is an impossible question, for if a real A gives birth to a real B, the reason for this change can be inquired into, or even found out satisfying the Leibniz's law of sufficient reason, but if real A pretends to be B without creating it, the question of finding reason for birth of B is an impossible question; for A and B are not of the same kind. One is the reality and the other is non existent like son of a barren woman or horns of a horse. Further the infinite consciousness is not really the Creator of this Universe, as one would expect it to be. The hot air, the trees or other objects, the refraction of light in a concave way with its convex side downwards, do not intend to create the mirage. It is the perceiving man whose sense organs including his mind that create the mirage for him. It is the mind, which creates Universe appearance even when there is no such intention in the infinite consciousness, which exists even when there is no perceiver or when the perceiver is in deep sleep state.

Physicists need not become dejected knowing this, for they are doing immense work in knowing 'How' of the apparent Universe, although now it is being realized by them that the role of the mind is vital in all perceptions in this Universe full of qualities.

Well, if starting from matter which is the sturdiest thing thought in this Universe, we move to time and space that are so consistently necessary to understand the Universe and we come to the conclusion that the Universe is nothing but an illusion, what should it mean really? If it is an illusion then all arguments are a part of this gigantic illusion, and hence they are unreal and thus the conclusion that the Universe is an illusion becomes wrong or itself illusive. Thus the Universe must be a reality, as it impresses upon us. But if it is a reality then the arguments that lead to declare it to be illusion must be termed as real, and they result in proving that the Universe is an illusion. This riddle is not as difficult to solve as it seems to be. If we say that the arguments are not real, we prove to ourselves that the Universe is an illusion by simply assuming this. If the Universe is assumed to be real then the arguments prove it to be illusion. So in both cases the proved fact remains that the Universe must be an illusion, for only this theory befits both lines of thinking. The mirage is an illusion having the bent rays as the reality at the base. The water in it is unreal when one knows how mirage occurs. But to all others who do not want to know the reality, the water in it is as real as they themselves are, and there is nothing wrong on their part if they endeavor to harness best this water for their benefit. A fish in such waters does not want to perish, and that is how the world goes on.

References

Introduction

1. Suppose c is the velocity of light, r the distance between two events and t is the time interval then c2t2-r2 represents the real time interval between them and if this is positive, the interval is time like and if negative, the interval is space like. Readers may refer any standard book on Theory of Relativity for details.

2. The Holy Bible, Authorized King James version, by the Bible meditation League, International headquarters, Columbus, Ohio, Chapter I verses 1 to 7

Chapter 1

1. A brief history of time by Stephen Hawking published in Great Britain by Bantan press pp 158, 159.

2. Maya in Physics by NC Panda published by Motilal Banarasidas publishers private limited, Delhi, India, reprint 1999p

3. The edge of science by Richard Morris, published by Fourth Estate ltd, 289 Westbourne Grove, London 1992 pp 95 to 101

4. John Wheeler in interview with Cosmic Search, Cosmic Search Volume I no 4, Forum John Wheeler, page 1 of pages 1 to 11 taken from internet

5. History of Western Philosophy, by Bertrand Russell, published by Routledge, 29 West 35 street New York NY 1001, reprinted 2002, pp 211, 212

6. History of Western Philosophy, by Dr Vatsyayan, published by Kedar Nath Ram Nath 132, RG College Road, Meerut (UP), India p 90

7.   Leibniz says that if there is no sufficient reason for a thing or statement to be what it is and not to be different from it, then its actual existence can not be real. He held that this law of Sufficient Reason is applicable to both in the field of metaphysics as well as logic. From Logic, By Ram Nath Sharma, published by Kedar Nath Ram Nath Co. Meerut(UP), India, First Edition, p34

8.   ABC of Relativity, By Bertrand Russell, reprinted in 1997, by Routledge II, New Fetter Lane, London EC4P 4 EE p25

Chapter 2

1.   The structure of time by Williams Newton-Smith published by Routledge and Kegan Pauls, London pages 83 to 87

2.   Test book titled 'Logic' by Ram Nath Sharma, published by Kedar Nath & co, Meerut (UP) India, first edition page 34

3.   The structure of time by Williams Newton-Smith published by Routledge and Kegan Pauls, London page 84

4.   John Wheeler in interview with Cosmic Search, Cosmic Search Volume I no 4, Forum John Wheeler, page 6 of pages 1 to 11 taken from internet

Chapter 6

1,2,3 - History of Western Philosophy, by Dr Vatsyayan, published by Kedar Nath Ram Nath 132, RG College Road, Meerut (UP), India p 189

4- The dancing Wu Li Masters by Gary Zukav, published by Bantam New Age Book, Bantam Books London, Bantam Edition 1980 pages 162, 163, 164

5-History of Western Philosophy, by Dr Vatsyayan, published by Kedar Nath Ram Nath 132, RG College Road, Meetut (UP), India p 189

6- History of Western Philosophy, by Dr Vatsyayan, published by Kedar Nath Ram Nath 132, RG College Road, Meetut (UP), India p 210

Chapter 7

1. A brief history of time by Stephen Hawking, published in Great Britain by Bantan press, a division of TransWorld Publishers, page 185

www.ingramcontent.com/pod-product-compliance
Lightning Source LLC
Chambersburg PA
CBHW032011170526
45157CB00002B/653